Language in the Brain

Linguistics, neurocognition and phenomenological psychology are fundamentally different fields of research. Helmut Schnelle's aim is to promote an interdisciplinary understanding of a new integrated field in which linguists will be competent in neurocognition and neuroscientists in structure linguistics. Consequently the first part of the book is a systematic introduction to the functional constitution of form and meaning organizing brain components. The essential core elements are perceptions, actions, attention, emotion and feeling. Their descriptions provide foundations for experience-based semantics and pragmatics. The second part is addressed to non-linguists and presents the structure foundations and formal presentations of currently established linguistic frameworks. This book should be serious reading for anyone interested in a comprehensive understanding of language, in which evolution, functional organization and hierarchies are explained by reference to brain architecture and dynamics.

HELMUT SCHNELLE graduated in 1957 with a degree in Physics. His Postgraduate studies between 1958 and 1962 included cybernetics, linguistics and philosophy, leading to the first doctorate in philosophy on Leibniz' *de Arte Combinatoria.* In 1967 he achieved his second doctorate (Dr. phil. habil.) based on the book manuscript *Prolegomena for formalization of levels of linguistics* and became Full Professor of linguistics in Berlin. He is a member of Academia Europea (London), Honorary Member of the Cercle Linguistique de Prague and in 2000 became an Honorary Doctor of the University of Bielefeld, Germany. He was Editor of the journal *Theoretical Linguistics* between 1974 and 2000. Helmut Schnelle also organized the first conference about language and the brain, on the occasion of awarding an honorary doctorate to Roman Jakobson and has since organized a "MIND/BRAIN" Conference in Paris. He is now continuing work with the Ruhr Universität Bochum, focusing on studying language in the brain and its organization of neural networks.

Language in the Brain

HELMUT SCHNELLE

Shaftesbury Road, Cambridge CB2 8EA, United Kingdom

One Liberty Plaza, 20th Floor, New York, NY 10006, USA

477 Williamstown Road, Port Melbourne, VIC 3207, Australia

314–321, 3rd Floor, Plot 3, Splendor Forum, Jasola District Centre, New Delhi – 110025, India

103 Penang Road, #05–06/07, Visioncrest Commercial, Singapore 238467

Cambridge University Press is part of Cambridge University Press & Assessment,
a department of the University of Cambridge.

We share the University's mission to contribute to society through the pursuit of
education, learning and research at the highest international levels of excellence.

www.cambridge.org
Information on this title: www.cambridge.org/9780521515498

First published 2010

A catalogue record for this publication is available from the British Library

Library of Congress Cataloging-in-Publication data
Schnelle, Helmut.
Language in the brain / Helmut Schnelle.
p. cm.
Includes bibliographical references and index.
ISBN 978-0-521-51549-8 (hardback) – ISBN 978-0-521-73971-9 (pbk.)
1. Neurolinguistics. 2. Cognitive neuroscience. I. Title.

QP399.S36 2010
612.8′233–dc22 2009052596

ISBN 978-0-521-51549-8 Hardback
ISBN 978-0-521-73971-9 Paperback

To Marlene

Contents

Preface

My ways to the studies of language were rather indirect. After having graduated in physics (1957) I read John von Neumann and Morgenstern's *Theory of Games* and was fascinated by their exemplification of the modern axiomatic method. I asked myself the burning question how far mathematical theories and formalization could lead in disciplines that develop beyond the natural sciences. The question led me to studies of the humanities. I first concentrated on the philosophy of Leibniz' *de Arte Combinatoria* and his *Characteristica universalis* and wrote my first dissertation thesis about *Symbolic Representations used in modern science* exemplifying, among other systems, networks of automata systems and the notations in Frege's *Conceptual Notation for Logic*. Subsequently I studied Neo-Humboldtian linguistics and wrote my second thesis about its possible formalization in terms of information flow networks.

An interesting research position about theoretical and computational linguistics and their possible applications to machine translation led me to many cooperation visits to research institutes in Europe, the United States and Israel, and participation at the 1964 International Colloquium for Algebraic Linguistics and Automata Theory about linguistic models in Jerusalem. During my years in Berlin I formally compared the theoretical varieties of Generative Grammar with the more mathematical models of Montague Grammar. Changing from Berlin to the new University in Bochum initiated a new start, caused by organizing a colloquium in honour of the famous linguist R. Jakobson at the occasion of his honorary doctorate. Since Jakobson knew that our group had already studied the clear introduction and detailed descriptions to functional brain architecture in Popper-Eccles' book *The Self and its Brain* he proposed the colloquium title: *Language and Brain*, hoping that we thus joined the new orientation he had described in a well-known New York University lecture. His words were "Progress in neurolinguistic research demands an even closer linguistic, or to put it more exhaustively, semiotic approach. The joint efforts of linguists and neurologists are summoned to suggest and open even deeper insights both into the structure of language with reference to the brain and into the structure of the brain with the help of language."

In the following years I constructed dynamic information flow models and dynamic organizing language structures thus competing with early connectionism. Here also I agreed with von Neumann's challenge that we need explanations of perception, action and thought based on *architecture and neural compositions of organisms*. Purely formal mechanisms or constructs like those of Turing or Chomsky are explicitly *limited to proving feasibility in principle* and thus not sufficient for understanding the complexity of human perception, thought and language. Turing did not contradict. In his self-criticism he even acknowledged that the precise notion of computation mechanisms *implies principled constraints*, which are relevant in various types of practical applications. He characterized one of them in an ironic mood: "Machines can't do certain things such as enjoying strawberries with cream. But the reason is not that computers and brains differ in operative architecture. Possibly a machine might be made to enjoy this delicious dish, but any attempt to make one do so would be idiotic." I believe with von Neumann that normally the architecture of our nervous system is made to enjoy strawberries with cream. Our difficulties in constructing corresponding machines should not prevent us from studying empirical neural organization that is relevant for generating this joy. I must say that the brain's organization of joyful self-experience is indeed interesting, and sections in my book will study present knowledge about these phenomena.

Parallel to these studies of Turing's self-critique my interest in complex phenomena was further encouraged by learning from my wife basic characteristics of creative invention and interpretation of visual art based on phenomenology and the neurocognitive details of visual thinking. Part of what I learned from the discussions or reading her books (1990 and 2002) is presented in particular sections of this book. Fortunately our common interests in neurocognition of vision and art arose at a time in which excellent new books were published and caused common studies over many years.

Let me now turn to the construction of the book. The discussion of language in the brain is confronted with three disciplines, studying language, studying the brain, both participating in phenomenological studies of mind. It is clear that for a comprehensive understanding of the same fact each discipline can contribute aspects that are appropriate in its own methodology, terminology and theoretical framework. In my view it is unfortunate that generally the disciplines remain separated. They shouldn't! On the contrary, comparison of interdisciplinary characteristics would lead to mutually fruitful conceptualization.

I think that the most appropriate ways to become acquainted with interdisciplinary correspondences is to present the book's content in two parts. The first introduces to linguists phenomenologically or mentally structured neurocognition; thus *functionally marking* brain architecture and processes. The second part concentrates on certain conceptually defined structures of grammar, lexicology, meaning and pragmatic usage of expressions or utterances. Here also many components are functional linguistics in the sense that phenomenological analysis is brought into correspondence with structure description.

Here is a brief survey of the four chapters of the *first part*. The first chapter explains the functional roles of neural networks in the brain. The neurons form internally organized clusters. Many narrow or distant clusters are mutually connected clusters that may exhibit mutually simultaneous activity patterns. In the case of *functionally characterized* smaller or larger cluster networks – for instance when they represent word or phrasal structures – one may say that their synchronized activation patterns represent momentarily active pieces of knowledge. Vice versa we may say that different pieces of linguistic competence knowledge exist in the brain as different distributed neural cluster activities. The related *pieces of knowledge* and functionally coherent clusters are functional units called *cognits*. This cluster network is mainly located in the cerebral cortex. It is moreover related to other parts of the body such as sensory perception or the muscles or the visceral system that organizes the internal body organs. The cortical clusters and the two other components are interdependent and interactive. The peripheral and internal connections exist already at birth, whereas the completion of cortical structure patterns takes many years through infancy, childhood and adolescence.

While the first chapter concentrated on principles of cluster networks in the cortical architecture the second chapter concentrates on their role in perception and action organization characteristics in the mammalian cortex. The chapter also introduces a radical extension of the perception–action hierarchy existing already in all mammals—somehow forming a concrete semantic organization. The *Homo sapiens* brain combines this system with language form that is *grammatically organized speech sound perception and action* located in *Broca's and Wernicke's areas.*

Both semantic and grammatically organized systems offer two kinds of operation: lower level *automatic self-organization* of normal and standard grammatical form and operations in which structure components can be *selected* and *composed in more complex ways.*

The third chapter will discuss a number of brain functions organized by perception–action systems whose measurements and detailed analyses marked breakthroughs for our understanding of neurocognition. There was the discovery of the *mirror neuron system* explaining the existence of functionally important interdependencies of perception and executive organization of actions. Measurements of *developmental characteristics contributed to the understanding of perception–action complexity*. Of particular importance were *studies of vision*. It turned out that the organization of *vision is much more complicated than is commonly assumed. Saccadic eye-movements* already play fundamental roles in simple identifications of objects and situations.

Whereas the facts studied in Chapter 3 are automatic in the sense that their execution does not involve consciousness Chapter 4 concentrates on combination and integration of intelligent thought and feeling. In order to introduce the essentials of their character the chapter begins with characterizing the *phenomenology of acts of creativity*, in art as well as in science. The neurocognitive analyses show that the *organizations in the prefrontal cortex* play a dominant role. They are for instance involved in operations of selection, attention, intentionality, thought integration, selective operations and evaluations of constructive thoughts and imaginations. These characteristics indeed play an important role in all *actions of creativity, whether in science or art* and also in constitutions of self-experience and the interpretation of and empathy with other self's experiences.

As a summary of the linguistic aspects of the book's first part we may emphasize that not only Broca's and Wernicke's areas of the brain contribute to language organization. They are rather involved in organizations of language form. Language meanings involving perception and action of concrete situations as well as feelings connected with them are organized in almost all components of the nervous system. This means that semantics and pragmatics have a completely different neurocognitive status than language form organization. Language studies that are reduced to formal language structure of syntax and formal semantics are too limited and inappropriate for practical usage of language. As in the case of Turing machines and formal rule-based syntax they may contribute to defining "structure feasibility" in principle. But what is needed is the understanding of concrete semantics and pragmatics or a neurocognitively interpreted semiotic approach, as Jakobson said. The explanations of the second part will show that language studies tend to develop in this direction.

The *second part* introduces linguistic approaches. It begins in the fifth chapter with explaining descriptive aspects, which strongly determined the second half of the last century. Carnap's earlier idea that syntax is the core of formal logic was accepted by the linguist N. Chomsky though in a radical adaptation to the conceptualization of the traditional grammar and inventions of linguistically transparent representations. Since this transparency is still valid, in particular for learners, the chapter begins with their exposition. But in the last decades many linguists followed the tendency to extend explanations to more integrated combinations of syntax and semantics. Chapter 5 continues by discussing essential stages of extensions proposed by the influential linguist R. Jackendoff. But when considering the range of neurocognitive knowledge presented in the first part of the book we are led to the conclusion that important parts of concrete semantics, in particular those characterizing attitudes and self-experiences of persons, remain inappropriately represented and characterized in terms of logically based conceptualizations. The end of the chapter criticizes this approach.

The approaches discussed in the sixth chapter follow a typical and influential variety of cognitive- and usage-based linguistics as it is presented in Langacker's cognitive grammar. Grammar is not understood formally or schematically but as a "grammar in life". The chapter discusses a selection of Langacker's proposals. They are partially based on phenomenologico-grammatical analyses. Specifications are provided by frames of distinct archetypes of understanding and by variations of flexibility of expression. They often contrast with Jackendoff's analyses that are represented in the fifth chapter in terms of formation rule generated structures.

Grammatical organization in Langacker's system can be understood as being based on different *examples of phenomenological and philosophical archetypes*. In addition to the archetypes there are considerations of the linguistic relevance of philosophical distinctions like *objectivity* and *subjectivity*. The different discussions lead to very interesting explanations of how the words are selected and grammatically arranged when a speaker intends to utter knowledge and grounds the structure by word and particle arrangements in the sentences. It should be obvious that our brief survey of semantic and pragmatic interplay can only name few of the elementary perspectives. The last two chapters invite the non-linguists to learn in which ways the different linguistic approaches distinguish linguists and their schools.

The last two chapters, Chapters 7 and 8, return to a central aspect of Jackendoff's *new perspectives for linguistic descriptions*, namely his idea to *push "the world" into the mind/brain/body* of a person. The idea per se is fruitful but requires the development of an *improved stage of analysis* leading to a *more radical reorganization* of the mind/brain/body analysis in its linguistic perspective. My way to the required schema will be prepared by fundamental critiques, revised explanations as well as new evaluations of selective powers of usage archetypes ending with final proposals for translations of formalist syntax into dynamic structure schemata that solve Jackendoff's critiques of connectionist proposals of cognitive neuroscience.

My essential critique concerns the constituent elements of sentence semantics, more specifically the fact that they are still used in the format that derived from logical abstraction schemata. They present sentence semantics in terms of the main verb that denotes *relations of objects*, an approach that leads to *argument structure models,* which are only secondarily sub-categorized. I still prefer approaches in which the persons, animals or dynamic objects denoted by noun-phrases have priority. On this base the semantic interpretations of body-based feeling, social attitudes and social group feelings and empathy can be accounted for. A particular exemplification will be presented in connection with the list of verbs that can be used in "each-other sentences".

The eighth chapter concentrates on further aspects of dynamic language organization. The first sections discuss the dynamic role of phenomenological interpretations of archetypical prototypes and background-based interpretations as well as very different types of efficient communicative function-based structures of sentences. The second part is rather technical in its studies demonstrating how formalist syntax structures are possibly translated into neural cluster systems. Using relatively simple syntax the translated result shows how configurations of syntactic trees like those that were introduced in Chapter 5 are translated in networks generating momentarily synchronized activity patterns corresponding to linguistically static descriptive configurations. It is shown how classic representation problems like token plurality and other difficulties can be solved. It is also demonstrated how cluster networks can organize context and background influences on syntactic alternatives.

Let me finally summarize the content. We are looking for new interpretations. Language in the brain should be described as a dynamic competence organization of form, meaning and usage. Neural networks are mentally activated in the intentional energy of speech acts and also historically changed by social *energeia*, as Humboldt said. These dynamic views should in principle be better substantiated by the analysis of *language in the brain* rather than *language in symbolic formalisms.*

Which are the essential elements of neurocognitive understanding? We should definitely not forget the fundamental integration of our brain organizations in the automatically self-organizing sub-systems, the perception–action organization system and the body's internal autonomous and somatic nerve systems and their integration organization in the prefrontal cortex and the nervous system centres of the hypothalamus and thalamus.

Though language form organization is only a section of the perception—action system it is closely connected with practically all other sub-systems that contribute in many ways to our understanding and self-understanding, conceptually focused in selective ways.

Combining our competence of language and thought in the brain the following is the most important point: Given the complexity of our bodies and minds and given our selectivity in consciously focusing and literally or ritually fixing what we consider as basic, we normally do not acknowledge that what we know consciously is necessarily only a skeletal system of what seems to exist 'here and there' but is supported by a much more detailed infinity of elements constituting the flow of 'now and then'.

Acknowledgements

As explained in the preface many years of my interdisciplinary research concentrated on perspectives, dimensions and facts that would contribute to an ultimately unified understanding of language, mind and brain. Thanks go to the authors of many books and articles mentioned and discussed in the various chapters of my book. I also owe debts of gratitude to many colleagues and friends who were ready to discuss my ideas in their details and their interdisciplinary perspective, in particular M. Arbib, M. Bierwisch, W. Brauer, P. Eisenberg, G. Fanselow, J.A. Feldman, A. Friederici, M. Gross, E. Hajicova, B. Johansson, M. Kay, S. Kanngießer, T. Kohonen, M. Krifka, W. J.M. Levelt, U. Maas, U. Mönnich, C.F. Küppers, R. Posner, G. Rickheit, G. Rizzolatti, P. Sgall, J. Sinclair, I. Wachsmuth, Chr. Von der Mahlsburg, W. Wahlster, and D. Wunderlich. I owe particular gratitude to the many discussions with my friend and teacher Y. Bar-Hillel, and the fruitful talks with V. Braitenberg and F. Pulvermüller, which led me to functional and concrete neuroscience. But much more influential for my interdisciplinary thought were the almost daily discussions with my wife, the historian of art and photographer, about the relations of neurocognition, phenomenology and gestalt psychology, in particular in their application to acts of production and perception of art and the possible in which our brains might organize creativity in art and science. There is no better way to express my gratitude to her involvement than to acknowledge that she was my worst critic, my best critic, and day to day source of inspiration and reason.

PART I

Functional neuroscience of language organization in the brain

1

The brain in functional perspective

1.1 The functional triangle of language, mind and brain

In a classical perspective a language consists of the set of its words and sentences determined by its lexicon and grammar. Words and sentences are realized as sound patterns and are mentally registered when we *hear them as sound patterns* or when we *identify them as letter figures on paper*. Today they can also be realized and identified as *letter configurations on the computer screen*.

But there is more. When *recalling* something said to us, the *memorized* words and sentences *appear as sound images in our minds* together with our mental understanding of the *words' and sentences' meanings*. We may also learn that, while our mind thinks, understands, or speaks, some of the grey cells in our brain are active. In a naïve understanding it may appear to us that pieces of uttered words and meanings are realized and kept in the brain like being printed on a physiological *tabula rasa* or in a storage space.

Many linguists disagree with the assumption that our mind images *everything* that is relevant for speaking or understanding. They emphasize that when we speak correct language we have no conscious image of all aspects of meaning and the rules that determine grammatical correctness. Indeed for speaking and understanding normal words and sentences the system of grammatical regularities is *somehow* operative, but we almost never have conscious mental images of them. Thus we must assume that the *rules of grammar* and *the rules of lexical word relations* can at best be represented structurally in the manner of an *abstract system description*. The mental system seems to be similar to other systems of rules such as the well learned intuitive competence of the rules of chess. When we are fluent players of chess we play without consciously concentrating at any moment on the rules. We rather master them spontaneously. Consequently grammarians conclude that in the linguistic system sense also is best understood as a spontaneously functioning formal system structure, a specific kind of mental entity that is different from the concrete images our conscious mind remarks. Let us call such an abstract system a *formally mental* entity. The theoretical linguists insist that the essential core characteristics of a language form are sufficiently conceptualized in terms of such formal systems. Consequently theoretically precise results of linguistic studies should be represented by functionally mental entities. In the theoretician's view the ideas that language is in the air, on paper, on the computer screen, in a computer internal data space, in images of our conscious thoughts or occur as activities of our brain cells may well be neglected.

I think that the formalist representations denoting formalized concepts have some advantage, for instance in presenting structure constructions in clear transparency. In this perspective formally structured language is recommendable. Acknowledging this does not, however, exclude developing and applying further mental perspectives of analysis that may appear to be more revealing for more comprehensive aspects and phenomena. There are for instance good reasons to carefully study the characteristics of *psychological and phenomenological phenomena* of situation-supported language use as well as the *complex organization that our brain contributes* to our knowledge and use of language. We thus should consider all three perspectives: (a) *Language* in the *formalist linguistic sense*, for instance in terms of formalized mental systems; (b) language in *verbal imaging and conscious phenomenological reflection* in our phenomenological mind organizing speech acts; and (c) language in the biological sense determined by *complex brain architecture* as a *complex brain activity*, as well as by biological development.

In fact, it can be shown that all of the three disciplines produce their precise analyses, each completely justified in its own methodological framework. On the other hand I am certain that mutual comparison of thoughts and models as well as combinations of perspectives can open new insights and direction in each domain. I thus recommend that precise analyses in these three domains should not be kept separate and isolated. Mere collections of methodologically separate studies will not lead automatically to comprehensive understanding of the perspectives of language organization. Instead the phenomena of each perspective must have their counterpart in each of the two other perspectives. We must even assume that understanding the phenomena in one perspective is improved when the characteristics are also *functionally* distinguished by specifying their role in the perspectives of the other frameworks. The following chapters will provide many indications of how brain architectures and processes distinguish potentially the organizational functions of grammar, meaning and pragmatic usage of languages. I insist that studying functional interdependency of interdisciplinary perspectives of language is very important and improves understanding of the general principles of language. Neither formalist linguistics nor phenomenological analyses of communicative intentionality and thought nor neurophysiological brain measurements are sufficient.

Figure 1.1 represents a triangle of disciplines whose functional interdependencies we should study and try to integrate. For each discipline we should also learn to differentiate phenomena relative to the roles they would play in the neighbour discipline. This openness would be possible after taking off the discipline's own blinkers. Clearly structured correlation of the three disciplines would generate *functional* disciplines, namely *functional neurobiology* (with respect to language structure and to phenomenology), *functional linguistics* (with respect to neurobiological brain organizations and to phenomenology) and *functional phenomenology* of intentional speech acts and thought analyses (with respect to linguistics and neurobiology of language organization). Their combination would create a *new understanding of functional cognition.*

I do not believe that this aim is utopian, but I am sure that reaching it is very difficult and the progress will require decades and centuries. Above all it is clear that a disciplined open mind and careful engagement of functional interdisciplinary analysis will be required because widely shared sceptical attitudes must be overcome. It is the aim of the present book to contribute pieces of understanding and supporting information leading to interdisciplinary studies about language structures in the mind and brain, about the brain's organizing language and mind and about the phenomenological analyses of speech acts and intentional thought.

Let me add an answer to a critical remark that might be advanced against my interdisciplinary triangle. Some linguists and philologists would think that

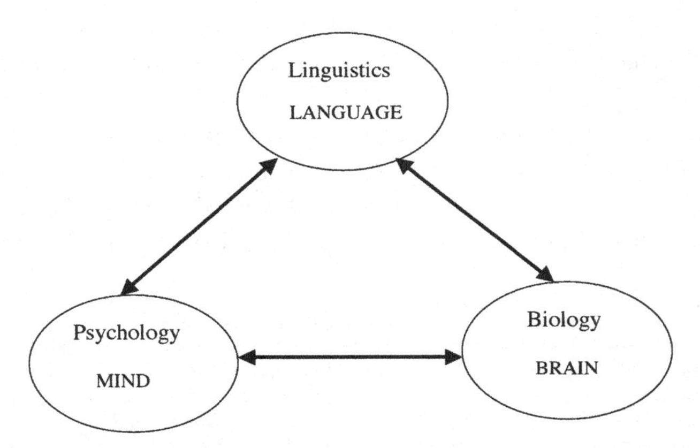

Figure 1.1. Triangle of functional interdependency of language-specifying disciplines.

the mind/brain analysis refers only to individuals' language knowledge in mind or brain and would consider such a view as inappropriate. They are persuaded that natural languages, such as English, German, French etc., must be understood as a kind of social institution; the individual merely participates and lives *in* this institution. In a sense I do not disagree. But in my view being a member of a social group is primarily the competence of applying implicit knowledge and assumptions about other people of the group one belongs to. A child acquires this knowledge during the first two decades of life. The adult person is able to put herself in the position of somebody else and only secondarily because she is formally obeying the rules or regularities of an institution. Each individual *has other mind knowledge and suppositions* that constitute the other person's social status. This other mind knowledge comprises indeed language competence of other speakers. Thus the collection of all speakers' varied individual competence of language and of its implicit communicative presuppositions about *other people and their common language usage* is a *sufficient base* for the institutional state of language.

Whoever agrees with this understanding based on the *other-self accounting perspective* might now be interested in learning the *principled characteristics and the details of the triangle of interdependent functionalism.* The following chapters and sections will present clarifications and explanations. I hope that they help to promote interdisciplinary research of functional linguistics and functional neuroscience by potentially supporting the other discipline's knowledge. But note: In the present research situation it is still impossible to present strictly established truths or established theories. *Baars' principle* is correct that science always makes inferences and assumptions that go beyond raw observations, using abstract concepts and descriptions that "make a believable story." Like any other start, ours also could encounter surprises. If so, we should be ready for changes wherever they are necessary or would lead to more plausible or more fruitful "stories."

With this insight we should conclude that universal claims of a discipline that tends to *substitute the triangle of functional studies by rigorous "unification of everything" in its own field* should be avoided, whether the unification base is formalist, phenomenological[1] or physically biologist. Though people

[1] The notion of phenomenology should be taken in the wide sense of *mentally critical psychology*. J.L. Austin introduced this perspective under the term of linguistic phenomenology in his article "A Plea for Excuses" p. 130 (in his Philosophical Papers 1961). In our context the most fruitful developments were presented and based on the notions of intentional acts by J. Searle's analyses (1969) (1983) (1992) and (1998), and on other levels Merleau-Ponty (2002) and certain aspects of Gestalt psychology. We will see below that the latter plays a certain role in the functional neuroscience of J. Fuster. These remarks were slightly extended to at least hint which aspects of psychology play a role in our context, though being less explicit in the details.

may have their own philosophy, as I have mine, all of them should acknowledge that the time has not come to rigorously argue for monistic positions against dualistic ones or vice versa or even bringing other philosophical positions into play. Philosophical discussions of this kind do not help methodologically and theoretically careful studies. Should we be able to attain a clear understanding of functional interdependencies as indicated by the triangle, it would very much help in the ultimate clarification of philosophical positions. Insisting on a strictly materialist understanding of brain processes does not really help. Why should we insist on eliminative materialism?

Instead the methodological and theoretical distinctions of the three disciplines and the openness for finding correspondences should be carefully heeded. Otherwise we run into misunderstandings. A typical case is the controversy of Chomsky and Mountcastle. Chomsky (2000) quotes statements of Mountcastle (1998): "Things mental, indeed minds, are emergent properties of the brains." Speaking of emergence does not help our appropriate perspectives about our selves. Successful clarification of perspectives prevents eliminative monism, whether materialist or formally mentalist. But Mountcastle insists. On the next page he claims: "All mental states are brain states." This is obviously misleading. A mathematician's or a linguist's concrete act of understanding in a framework of complex knowledge content is *a phenomenological analysable* act *of complex thought*, which in turn *corresponds to* a more or less extended *process of brain states*. In my view a correspondence exists indeed even when logicians and theoretical linguists follow Frege in believing that a thought's *content* itself is sufficiently understood as *an abstract structure entity*. But I insist that a *complete understanding implies the correspondence of several types of levels*: abstract structures, phenomenological knowledge processes and brain dynamics. All three contribute to cognition that will one day be completed by unification of correspondence understanding.

In our normal understanding each of the different perspectives provides a strictly different status of *cognition*. I think that my revision of Mountcastle's forced unification could lead to more careful phrasing. I fully agree with Chomsky's conclusive statement: "A primary goal is to bring the bodies of doctrine concerning language into closer relation with those 'emerging' from the brain sciences and other perspectives. We may anticipate that richer bodies of doctrine will interact, setting significant conditions from one level of analysis for another, perhaps ultimately converging in true unification. But we should not mistake truism for substantive theses, and there is no place for dogmatism as to how the issues might move toward resolution. We know far too little for that, and the history of modern science teaches us lessons that I think should not be ignored" (Chomsky 2000, 27).

On the other hand Baars' principle should not be forgotten: In the present research situation it is still impossible to present strictly established truths. I agree with the principle that science always makes inferences and assumptions that go beyond raw observations, using abstract concepts and descriptions that "make a believable story". Like any other simplified story, ours could also encounter surprises. If so, we will have to change it accordingly.

A last remark about my book's structure: It aims to support perspectives of research in the functional triangle unit. The chapters' arrangements are determined by didactic considerations. The reader should be able to learn about the brain and the language. The first part of the book is intended to inform linguists and students of the humanities and of philosophy about the brain. In the second part, discussing the relation of language structure, meaning, and development as a component of mind will introduce the neuroscientists to some core characteristics of present linguistics. The last chapter of the book will be more "technical" and present constructions of some of my working models. Formalist structure representations of language are translated into possible counterparts in neural network representations. Given the more formal aspects of this part it is unavoidable that some passages of this chapter will be a challenge for readers who are not familiar with formalist descriptions.

1.2 Introduction to the brain: the cortical network elements

Since Galen, antiquity believed that nerves are ducts conveying fluids that are secreted by the brain and the spinal cord and transmitted to the peripheral locations of the body. But it was completely unknown how the central brain tissue operates. Was it a continuous reticulum, a tangle of netlike biological tissue? The invention of the compound microscope in the eighteenth century showed that the cortex was indeed a tissue, though like a structure of mixed stalks; somehow similar to Figure 1.2.

By the end of the nineteenth century the neurologists Wernicke, Sherrington and Ramón y Cajal introduced a new and empirically based theory according to which the brain's function found its base in a cellular system of *connectionism*. According to this view, individual neurons are the signalling units of the brain. They are generally arranged in functional groups and connect to one another in a precise fashion. This view became basically influential until recently. Fortunately the first half of the last century brought much empirical and conceptual progress. It led Hebb (1949) to propose a model whose simplest components were presented as in our Figure 1.3.

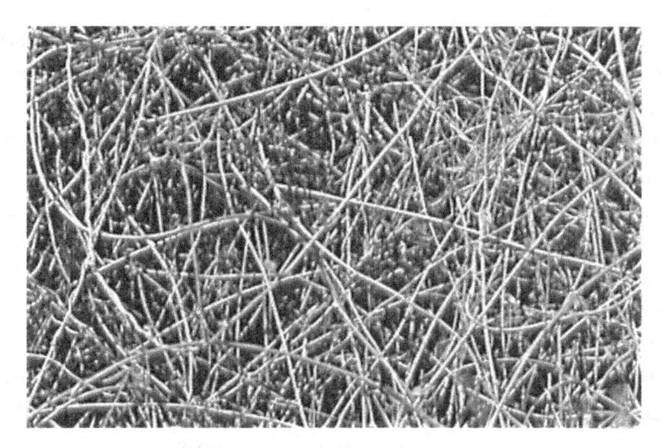

Figure 1.2. Is the forebrain a confusing tangle?

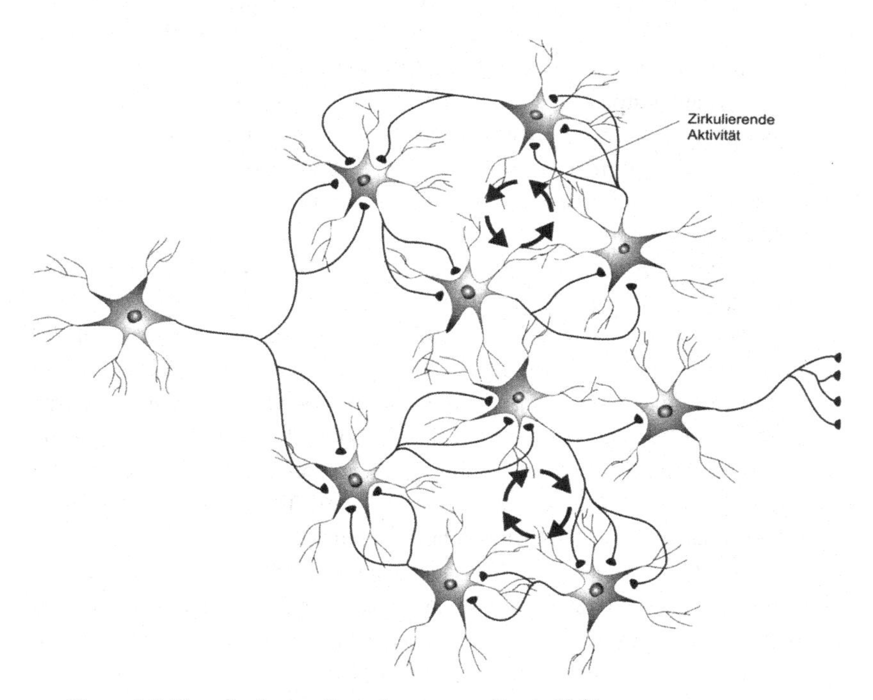

Figure 1.3. Neural sub-structure schema according to Hebb.

The cellular connections operate in very different areas of the cortex. The neuroscientist Kandel (1995, 8) wrote that the brain functions in the cerebral hemispheres are like the *bark on a tree*. In each of the brain's hemispheres the overlying cortex is divided into four anatomically *distinct lobes*: frontal, parietal, occipital and temporal. They contribute to different sub-aspects of more globally organized specific functions such as the planning for future action, the execution and control of movement, tactile sensation, body image, hearing, seeing, learning, memory and emotion. Thus different brain components organize a single behaviour and cooperate in different regions of the brain.

However, towards the middle of the last century there was growing scepticism concerning this model. Particularly influential was Lashley (1950), who claimed that the organization of specific functions resulted from non-localized mass operation of cellular connectionism, a position that influenced fundamentally Chomsky's linguistic view. Lashley argued, and Chomsky agreed, that learning and other mental functions, such as advanced linguistic competence of language form, have no special locus in the brain and consequently cannot be related to linguistically relevant networks of neurons (Kandel 1995, p. 15).

But subsequent discoveries of neurocognitive science, based on mainly microelectrode measurements and techniques of brain imaging, provided sufficient evidence against Lashley and Chomsky's radical scepticism. They again justified the Sherrington and Ramón y Cajal idea of cellular connectionism, now, however, in a new form. Instead of single neuron connectionism, *complex groups of hundreds of neurons are the functionally operative units* each performing rather elementary specific operations over the network of mutual connections. In any case it is clear that these considerations made the step from *global interaction of cortical areas* to *microscopic analyses*. They provided a first understanding of the complexity of dynamic units. Their systematic analysis and other studies by Mountcastle and Hubel and Wiesel led Szentagothai and Arbib relatively early (1975) to the stereographic view of the neuron cluster represented in Figure 1.4. It conveys the state of knowledge at the end of the 1970s. The problem with the figure is that it looks like a single cluster being arranged around an arborization of a single cortico-cortical *afferent*. The careful reader would also remark about the efferent axons. The afferent and efferent connections characterize the cluster as being an element from a widely distributed network. But the reader should conceive a neural arrangement that is more in accord with a more modern understanding. The proper arrangement contains many intra-level excitatory and inhibitory neural activity connections to neighbouring clusters. They determine a neighbour connection system that shows that there is no conflict whatever between cluster continuity or discontinuity.[2]

The interaction network of such modules contributes specifically to efficient cooperation, which realizes complex experience and behaviour and elementary perceptions of sound feature arrangements.

[2] About more detailed explanations, see these systems in M. Arbib et al. (1998).

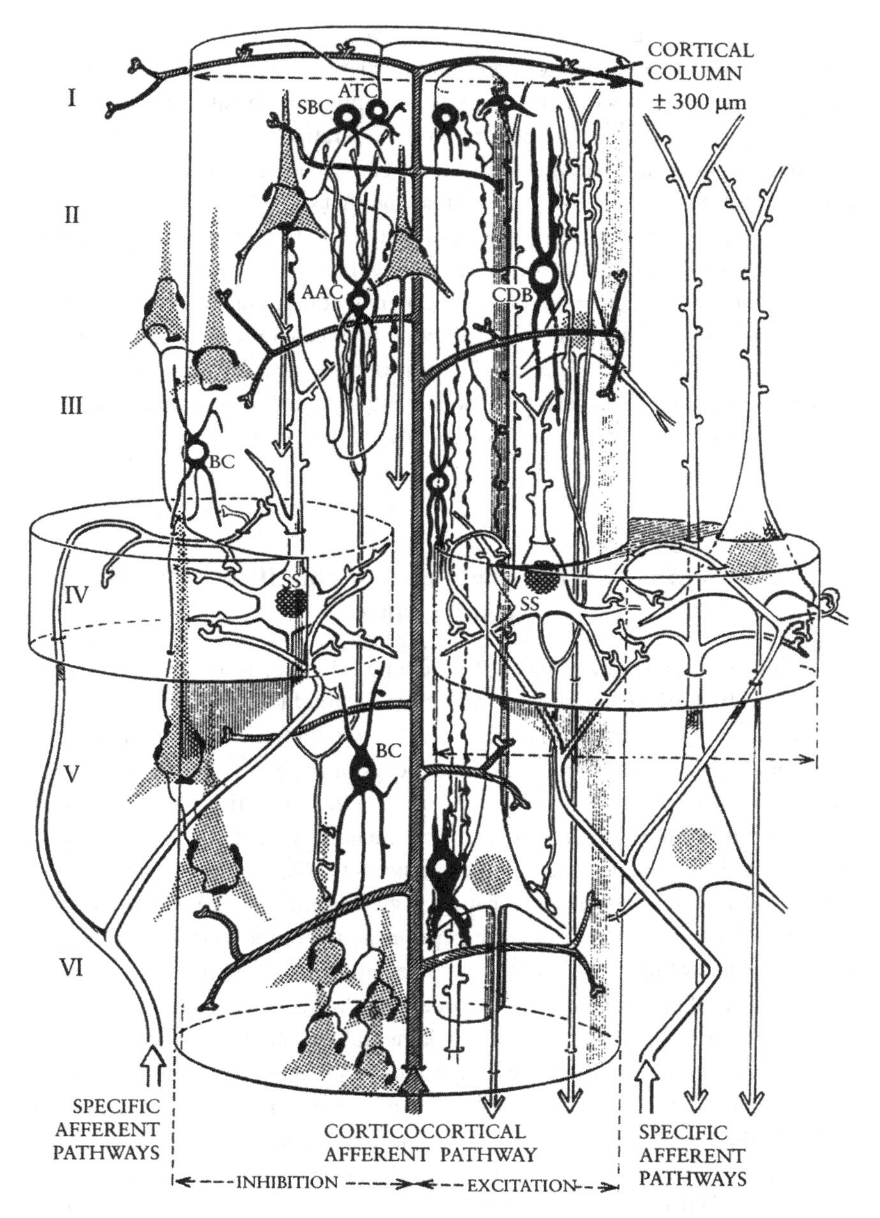

Figure 1.4. A schematic neural network contained in a local cluster module.

Parallel studies have demonstrated the developmental condition according to which the neurons result from pre- and postnatal cellular growth in the cortex. Very different types of neurons move to appropriate positions in six layers in the 2-mm-thick cortex. Depending on the operative roles of different areas there are neighbouring arrangements composed of larger clusters of 5–10 mm and smaller clusters of 5–1000 mm. It seems that the evolution of mammals determined the general principle of these vertical and horizontal arrangements in allowing the generation of the basic frames for their species-specific functional efficiency. Specific types of environmental conditions influence the postnatal growth of specific cluster connectivity, thus generating various forms of fine-tuning of the cluster structure. The appropriate growth of the internal and external connections generates the fundamental elements for individually and socially appropriate behavioural efficiency of behaviour and thought and feeling.

1.3 Cognits and the principles of cognitive network organization

Analysing and interpreting interactive cooperation leads us to suggestions about how *biological activity* of the networks might *correspond to mental interpretations*. It is appropriate to interpret *functional neural networks as pieces of organization* that correspond to pieces of the organism's mental competences, that is *pieces of knowledge*, or *cognitio* in Latin. Now, pieces of knowledge about things, events or situations are usually called mental *categories* or *characteristics*. Hence the corresponding networks could be functionally called *cognitive networks* and their mentally corresponding knowledge pieces are categories or characteristics of perception, action or thought. The pairing of a network and its functional knowledge piece might be given the name *cognit*,[3] but let us pay attention. Fixing names does not yet avoid mental and organizational confusions. There is still a fundamental problem for mutual understanding of pairs of linguistic categories and operative neurocognitive analysable networks. This is the main topic of the present book. It is important to clarify the essentials of the contrasts and pay attention to possible confusions as early as possible. The following explanations will concentrate on basic aspects only. Further details will be presented in the more specific chapters about theorizing in neurocognition and linguistics. In a few key words we may say: The difference derives from the basically *static interpretation of linguistic*

[3] As J. Fuster (2003) p. 14 proposed.

entities such as features, categories and grammatical structures arranged in configurations. But the neural networks that are category appropriate organize production and understanding of language utterances generated by their dynamic network power that activates appropriate action states given intentional conditions. The standard forms of linguistic, logic or other symbolic systems of abstract conceptual schemata thus eliminate or neglect the *fundamental role* of notions of *dynamic time*. This perspective is inappropriate since neurocognitive models rely necessarily on *dynamic and causal interaction models*. Let me give a brief account of some fundamental details.

First we must understand that essential parts of modern linguistics belong to the family of *knowledge analyses defined by formal or computational theories*, the two versions of formally symbolic theorizing. The main focus of interest is the clear *expression of precise order* in terms of symbols and strings of symbols that are explicitly generated or constructed by combinational rule operations. There is obviously a deep gap between cognitive configuration static and biological network dynamics. The former are understood as pieces of knowledge that are formally represented as static *linguistic configuration structures*.

Instead, language in the brain relies on networks of local clustered and connected neurons. Typically, elementary knowledge pieces, such as features and simple categories, are neurocognitively represented by simple and local clusters of dynamic type organization. More complex knowledge, such as structures of words, phrases and sentences are neurocognitively represented by distributed and interactive networks of cluster modules. *Interactively activated* and *synchronized* and situation appropriately "*energized*" *sub-networks* generate, given an appropriate context situation, a complex knowledge pattern. Thus, in the dynamic perspective, patterns activate momentarily relevant pieces of structured knowledge. It is obviously important that the gap between the static and the dynamic perspective must be overcome. We must understand the difference between the two types of representation: In the Greek sense of the word dynamic a "*dynamic* unit complex" means that its elementary units have the power of generating a *possible* operation competence, a power that may "wait" for an appropriate moment of activation. At this moment the cognit complex changes into the momentarily actual *energetic* state of *executive operation*. It is caused by internal interactions of the neurons in the network. The principle of organization should be clear, though in a superficially perceived arrangement the confusing complexity represented in Figure 1.2 remains.

Though most of us tend to believe that our thoughts and our speech intentions are organized in the brain even linguists now agree that there is no reason to assume that letters, bit patterns or category terms are stored in the brain and that some central processor accesses the symbol- or bit patterns and identifies, combines, separates or otherwise manipulates them in some working store area. The gap between static symbol manipulations and the self-active network units that realize energetic activity is obvious.

Both representational "mechanisms" must be better understood before we are able to bridge the gap between static and dynamic representations. Neither mere configurations nor connectivity networks indicate by themselves how they instantiate linguistically well-known formalist representations or statements. Even most experimental measurements show that neither microscopic nor macroscopic brain connectivity or architecture directly indicate the functions that neurons, neural clusters and cortical areas play in the organism's life. But I believe that the previous explanations about neurocognitive networks may be sufficient in indicating that the functional analyses of lesions and functionally determined microelectrode measurements and brain imaging are biological phenomena that could contribute to clarifying our understanding of functional distinctions and differences between the principles of linguistic and neurocognitive modelling. The models in turn should help to overcome the gap by translating symbolic frameworks into dynamic frameworks, hopefully indicating ultimately a kind of homomorphism.

But despite conceptual differences of modelling and theorizing, at least one linguist, namely Roman Jakobson, indicated a direction in which we should try to consider first *elements for bridging language description and neurocognition.* A brief account of his earlier research situation may be suggestive even today. His first intellectual stimuli were provided by empirical discoveries in studies of visual perception by Hubel and Wiesel (1962) and of tactile perception by Kaas and Merzenich.[4] They demonstrated the existence of arrays representing elementary cognit clusters (columns or modules) for vision and for touch. These studies were indeed very popular in the 1970s. Interdisciplinary discussions of the linguist Jakobson and the neurologist Teuber pondered the question of whether and how interactive columnar networks in the brain could represent distinctive phonological features and could "compute" activities for momentary feature patterns and combinations of linguistic sounds.

A brief outline of the early ideas may help understand the direction of possible translations as kinds of bridging. Jakobson and Teuber were stimulated by the empirical results of neuroscience published in the heydays of explaining structure and functioning of elements of cortical architecture. By the way, I must confess that the same aspects fascinated me during these years (Schnelle 1980c, 1981d). Teuber and Jakobson concluded that the following

[4] Compare the figures and descriptions in E. Kandel (1995) p. 337 and 443.

statement – which I would like to call the *Jakobson–Teuber principle* – could express their common view:

> The distinctive features [of phonology] would be more than a universal schema for classifying phonemes in all their diversity across languages; the features would be 'real' in the sense of being universal neural mechanisms for producing and for perceiving sounds of speech. (Jakobson & Waugh 1979, 123)

The most provocative term in this statement is obviously the word "real". What did Teuber and Jakobson mean? The crucial idea is that both were influenced by the notion of neuronal clusters discussed in their time: The neuronal columnar module cluster consisting of more than hundreds of neurons specifically connected and functioning in different ways. The discovery and explanation by Mountcastle in the years 1957 to 1967 (Mountcastle 1998) became very influential after the publication of very successful microcircuit studies. My own research during recent years led me to generalize this statement into the following generalized Jakobson–Teuber principle:

> The distinctive features of phonology, the linguistic unit denotations, and the linguistic categories on other levels are no longer understood as symbolic elements of universal notional schemata for classifying phonemes, or structuring words and sentences in all their diversity across languages. Instead, linguistic features, units and categories are to be considered as "real" in the sense that each is represented by universal interacting neural mechanisms, contributing to organizing dynamic patterns in the brain's cortex, understood as generating and controlling form, meaning, and context interdependency of all elements of speech perception and production. (The solution of problems that still remain will be solved in Chapter 8)

Obviously, the term neural mechanism is to be understood as the biological counterpart of a cognit, capable of generating the activity binding the linguistic features units, and categories, that is of linguistic cognits. More complex combinations of features or categories require more complex modules or module combinations integrated by complex binding processes.

The Jakobson–Teuber considerations thus proposed the first idea of a bridge between neuroscience and linguistics. They implicitly introduced the idea of a neural realization of linguistic-form cognits (LF-cognits) an idea that is easily generalized to grammatical features categories and structure-patterns.

The Jakobson–Teuber principles require the solution of some additional problems. Their idea would suggest that a distributed complex of local neural clusters should realize such a configuration. A remaining problem is *how the reality integrates the elementary cognits into complex understanding units*. In the neurocognitive models, which will be explained in subsequent chapters, it is at least clear that there are *two types of integration*. First there is integration by *automatic self-organizing* that is non-consciously generated, as happens in the case of not too complicated syntax organization. Second there are *symbol guided and feeling accompanied organizations of images and imaginations* well known from conscious and creative thought. These types of processing are well known in productive works of art and music as well as in their careful perception and in interpretation of pictures. They are also known in reports of mathematicians about their acts of passionate logical or mathematical reasoning. It is clear that these *types of integration* access complex automatic interactions of neural sub-components that exist already as partially non-conscious and partially consciously focused components of the nervous system's activity. These aspects will also be discussed in Chapter 8. The present chapter will instead continue to discuss general and basic aspects of neuroscience and neurocognition.

1.4 Mutual functionality: relating abstract linguistic structure, mind's phenomenology and functional brain organization

Despite the topographic muddles of neural networks discussed in the previous sections one may hope that clarification of specific functions studied in linguistics, psychology and phenomenology may help to develop understanding of the functional side of analysis, helping to disentangle the complexity of distributions in the brain, at least by reference to *working models*. Several decades of neuroscience have determined how *specific functions* can be assigned to complex interactions of brain areas: perception, action, language form, attention, imaginations, knowledge about the world, emotions, self-feeling and mentalizing other selves, social regularities of social behaviour, practical planning and controlling action execution, internal and external body feeling, memory storage, memory recall, memory-based thought processes etc. The most important function-determined brain feature is *mental interdependency expressed by neural clusters' interaction* in the brain. But there is a basic problem: There is no one-to-one mapping of mental or psychological functions assigned to separate brain areas. Instead the correspondence mapping of functions and locally coherent operative brain areas is many to many: Each function correlates with many locally coherent areas, and a locally coherent area usually contributes to the organization of several functions. *Broca's and Wernicke's areas* are no exceptions. *Each area correlates with just a sub-function of linguistic form.* When we consider complete language competence even these two sub-function arrangements are only parts of the organization of meaningful language competence. This suggests already the following basic statement: The mental function of *language competence is instantiated by many brain areas*. It is important not to forget the central neural function, namely interaction. In most cases only interaction of several functional areas allows them to generate the functional coherence of interdependent components.

Let us turn to clarifying the status of mental functions comprising those that were enumerated above. I hope that you understood, at least globally, the descriptive meaning of each of the functions mentioned above, namely perception, action execution, attention, emotion, language and intelligence. They are not strictly part of folk-psychology; the cognitive literature of the last decades has contributed to some clarification. Systematic reflections of epistemological analysis, of analytic philosophy, phenomenological philosophy, gestalt psychology, intentional and integrative frameworks of speech act theory, some varieties of psychotherapy, anthropology and human ethology, and in some sense, comparisons with behavioural studies of primates have contributed. Methodologically these analyses transcend mere analyses of measurements of people's external behaviour. In all cases what can be observed or measured requires careful reflection about underlying processes of mind, differentiated and systematically interpreted on the basis of personal experience. From a principled point of view, the most *careful analyses* seem to result from *phenomenological reflection* in philosophy, and from *fundamental studies of Gestalt psychology*. From a practical point of view careful considerations of neurologists often come to persuasive insights corresponding to relevant studies in phenomenology and psychology.[5] The following chapters will often refer to their thoughts and proposals. A more specific summary of the aspects with which I agree will be presented in section 4.7.

Let me briefly mention a methodological core feature, which I want to address to philosophers: A phenomenology that merely refers to experience in the framework of conscious thoughts of adults, as philosophers usually presuppose, is problematic. Our mental experience develops during infancy and childhood and remains related to foundational inner experiences that often dominate in childhood. A philosophy that does not account for these conditions of knowledge emergence and does not integrate the emerging

[5] For instance A. Damasio (1999), D.N. Stern (2000) and J. Bruner (1990). Particular extensive reference to Gestalt theory is made by Fuster (2003), emphasizing that Gestalt phenomenology is a particularly appealing method for the cognitive neuroscientist, a point which I strongly support.

archetypes of feeling with those of adulthood is highly problematic. Some experts of psycho-therapy, not only those who follow Freud, are particularly near to this idea.

I feel justified in this address. Neglecting emergent structures and archetypes is a typical bias of most philosophers. It is certainly characteristic for the original inventors of phenomenology like Husserl and Heidegger on the one side and the formalist analytical philosophers on the other. At least the former have shown that aspects of language meaning require clarification in terms of concrete accounts and careful phenomenological analyses of mental experience, showing that what they say cannot be reduced to descriptions of pure conceptual frameworks as formalists believe. Nevertheless a proper selection of mental reflections drawn from philosophical, psychological and linguistic knowledge, involving experiences, conceptualizations, intentionality and integrative mental aspects of perception–action organization and feeling will probably help the functionality details of our discipline triangle, namely language, mind and brain, and contribute in a methodologically fruitful development to an integrative framework of language understanding.

1.5 Introduction to the nervous system and its functions

No doubt, there is a growing agreement among linguists about the basic difference between grammatical and formal semantic and mental experiences of concrete perception and action. The former can be presented by conceptualizations as well as by categorizations of words' and sentences' form and meaning. The latter may relate either to immediate or memorized images of perception and action, or to concrete conceptualization of perception and action. The distinction corresponds to the fact that essentially different processes require different brain areas, as already indicated in section 1.2.

At the same time linguists usually insist that, although a great deal is known about functional localizations of various aspects of language in the brain, nothing at all is known about how neurons instantiate the organization of rules of grammars or how they store and retrieve lexical structures and meanings (Jackendoff 2002, 58). I shall try to show how our *understanding can be improved* and how basic *explanatory progress can be made*.

Given the more extended aims of this book, some basic knowledge about functional brain structure is indispensable, or at least basically helpful. It is definitely not sufficient to study the wrinkled surface of the cortex marked by its clefts and ridges. Superficial topography has only secondary relevance. Instead, we must start with considering the *interdependent systems of functions* and then turn to the task of a *functional explanation* of brain architecture, whose dynamic and energetic processes realize the functions.

As a global distinction we may present two basic functions, namely *cognition*, comprising perception, action and thought organization, and *body state feeling*, comprising emotions and personally felt self-experiences. Biologically understood, competences of *cognition* are mainly organized by the brain's *cortex*, whereas body organization of *homeostasis* as well as signals of body states and their configurations are organized in the *body distribution of the nervous system*. In certain states they cause partially experienced emotions and contribute to generate body-based *feelings of self-presence*. The two systems of the cortex and the body distribution of the nervous system are *anatomically separate* but interconnected functionally and at certain boundary areas. They *combine* the representation of *cognition* with the *sub-conscious self-experiencing cognizing*. As an example we may say that the *concrete semantics* of a sentence like "I feel cold" is not represented in the cortex but rather in the hypothalamic organization located in the nervous system.

In a general characteristic it may be said that the brain's components organize the following functions.: (A) The *brain's posterior and frontal cortex* realize the comprehensive *perception–action* system. Note that a component of this system serves, among other tasks, also the *automatic organization of phonetic, syntax, lexical semantics and imaging*. (B) The prefrontal cortex (PFC) organizes the *selective and executive* system. (C) The autonomic nervous system based on body distributed "large distance connections" that serve the *viscera*, the *smooth muscles* of the inner body as well as information about the *skin, muscles* and *joints*. These information data are centrally integrated in the *thalamus* and *hypothalamus* (T + H).[6] These data are finally connected to other systems determining emotion and motivation[7] thus generating feeling organization.

Let me now turn to some details of Figure 1.5 showing the distribution of brain components involved in organizing the functions. As just mentioned cognition and feeling are based on *integrated realization* processes of the *cortex* and the *body oriented nervous system*. This integration is mainly organized by the so-called pre-frontal cortex in the frontal part of the circumscribed brain complex in Figure 1.5. The dotted boxes at the cortical periphery line point to *the four lobes of the cortex*, the frontal, the parietal, the occipital, and the temporal. As will be explained later, activity in the cortical lobes organizes the functions of perception, action, cognition and thought. Core elements of emotion and feeling are organized by components indicated by boxes without dots,

[6] See further detail in W.J.H. Nauta and M. Feirtag (1986) Part II.
[7] cp. E. Kandel et al. (1995) Chapters 32 and 33. Also Moscovitch et al. (2007).

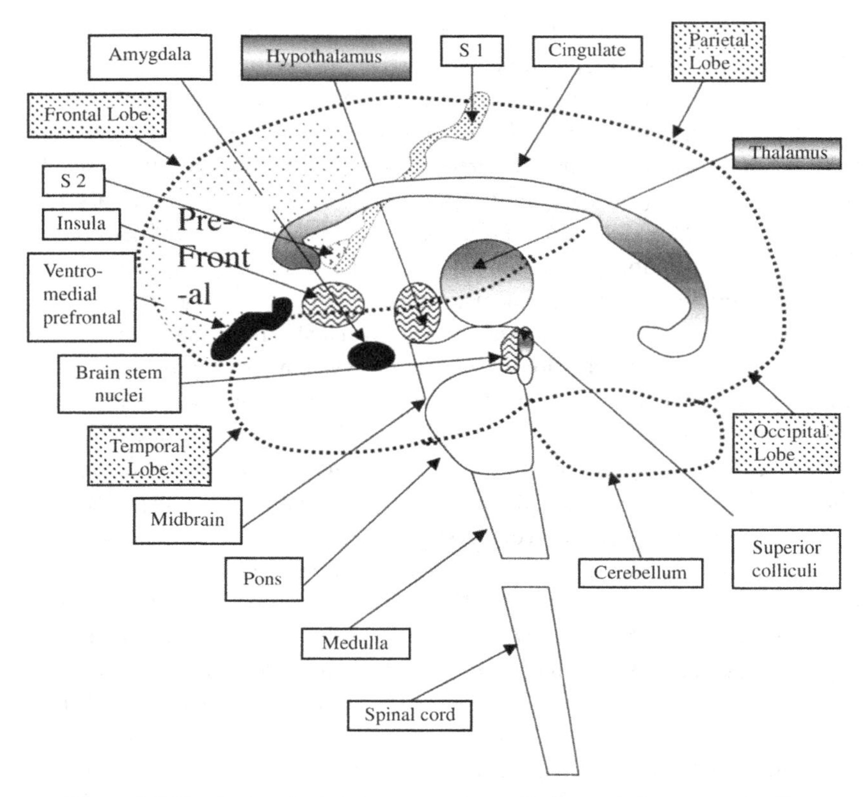

Figure 1.5. The body-oriented nervous system with its central components, the thalamus and hypothalamus.

for instance the white boxes, and the important components marked as *thalamus* and *hypothalamus*. Most of them constitute the so-called *limbic system*, for instance the *amygdala* and the *hippocampal system.* Together with the cingulate, the insula and somatic areas SI and S2 the pre-frontal cortex binds cognition and feeling organisation systems.

We now turn to Figure 1.6, whose pointed circumscribed complex corresponds to that of Figure 1.5, now intended to circumscribe the *left hemisphere of the cortex*, special areas of particular relevance for us: Broca's (B) and Wernicke's (W) areas for language form organization, peripheric cortical information organization of vision (V) and audition (centrally located A), attention organization (at prefrontal A), emotion connection (E), memory organization (Mem) and the peripheric motor organizations (M). The curved lines in Figure 1.6 represent left hemispheric fasciculi connecting Broca's and Wernicke's areas and other complexes in the posterior and frontal cortices.

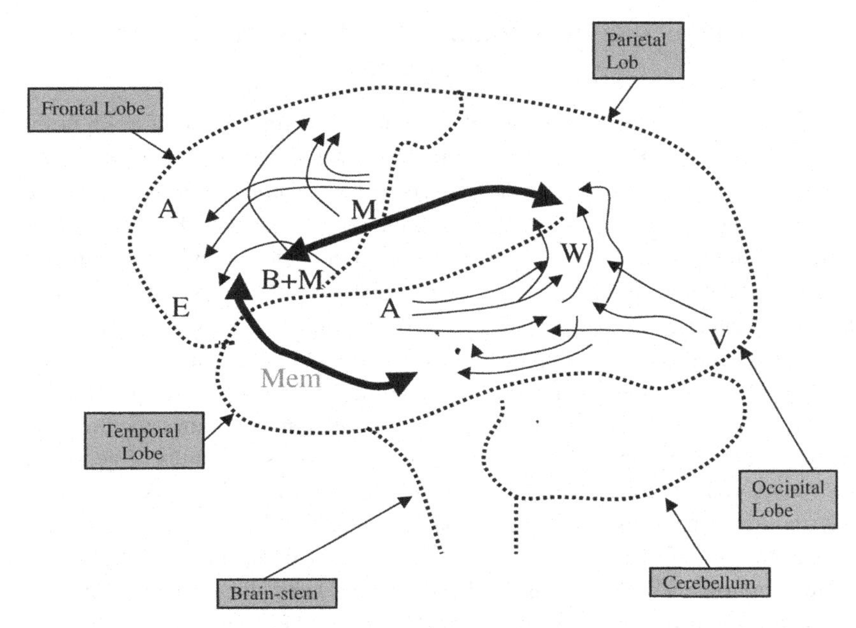

Figure 1.6. Cortical connections of Wernicke's and Broca's areas.

Whereas many units of Figure 1.5 are sub-cortical Figure 1.6 emphasises the four cortical lobes of the left hemisphere, the frontal, the parietal, the occipital and the temporal, in which special areas are located: Broca's (B) and Wernicke's (W) areas for language form organization, peripheril cortical information, organization of vision(V) and audition (centrally located A), attention organization (at prefrontal A), emotion connection (E), memory organization (Mem) and the peripheril motor organizations (M). The curved lines in Figure 1.6 represent left hemispheric fasciculi connecting Broca's and Wernicke's areas and other complexes in the posterior and frontal cortices.

The main part of the cerebral cortex will be discussed in the next chapter, demonstrating automatic perception–action systems independent of language, perception–action systems for language form, formal meaning, and concrete meanings of situation, memory of perception, action, and ontogenetic development. The organization and the importance of autonomic systems will be discussed in Chapter 4.[8]

[8] More details in T.L. Powley (1999) pp. 1027–1049, 1022.

1.6 Principles determining the development of the nervous system

Developmental stages of brain architecture mark the realization of brain functions. Each stage can be measured in months or years, mostly showing some variations of typical times. These aspects are *particularly interesting in the case of language development.* Which are the typical stages and growth conditions of language competence development? For their discussion we should consider some bottom-up aspects. The operative core elements of the complete nerve system are the nerve cells, their interaction via activation potentials mutually exchanged via synapses and axons as contact elements.

Already at birth the brain comprises hundreds of billions of neural cells and a basic network system of cells and connections, the innate brain structure. The innate system develops in the embryo; during the last six months of gestation the neural cells migrate in various areas of the brain in order to generate a rich distribution of neurons. Early growth processes of cell proliferation and migration arrange the *innate brain architecture* comprising different *types of neurons*[9] in neighbouring cell *clusters* also called *columns*. During the next stage neurons develop growth of *synapses* and synaptic contacts between neurons, thereby generating rich systems of *cluster internal neural connection networks.*

In addition to the growth of internal cluster structure, internal cluster external connectivity networks connect each cluster complex with more distant cluster complexes. Types of network combinations develop. In this process *fasciculi of long axons* grow to connect groups of clusters in *connection directions* that are functionally relevant. Some are cortico-cortical connections; others connect the clusters with cortico-thalamic and other autonomic system components. All of them provide widely integrating networks allowing relatively direct connection of every cluster group with any other group (see Petrides & Pandya 2002). Figure 1.6 represents some fasciculi, among them the fasciculi connecting *Broca's and Wernicke's* areas, two stronger and many thinner.

Just as other do neural connections the thinner and thicker fasciculi develop after birth under genetic control and thus produce distributed cortical connections, which evolution has determined as very important for the species. The *global frames of brain architecture* based on cell cluster arrangements and on short and long distant connectivity contain already some function-specific operative parts of the neonate's brain. These frames are fundamental for the

[9] Cp. N.M. Gage and M.H. Johnson (2007), Fig. 15.14.

neural growth processes of subsequent years, particularly concerning the cortex and some of its connections with sub-cortical units.

Only few functions are already operative a few days after birth: the eye-orientation directed to a human face, usually the mother, the mother's odour, the feeling of being hugged, hearing the calm rhythm of spoken motherese and so on. These are *biological imprints* in the human brain organization. But at birth the innate nervous system is still merely a "*skeletal*" *network arrangement* that must wait for further growth. But it provides, global as it is, a basic frame for guiding more differentiating development. Detailed reactions pass many stages of organization and even several processes of re-organization, as we all know from observing our children in their stages of development.

During the first year of life, the specific connectivity in the human brain changes in particularly dramatic ways. Shortly after birth the number of connections between the brain cells starts to increase rapidly. Due to this growth the number of cells greatly exceeds neural network changes in adults. Many of these *cells cut back* in subsequent stages of development. The reason is that their contingent patterns are inappropriate for the specifics of task organizations. The cutting back (often called pruning) is the initial part of the brain's adaptation to environmental conditions, resulting in *fine-tuning* of the brain's organization competences.

Once again, the growth of networks that will be able to organize the specific functions involves the following features: (1) The number of neurons at birth is sufficient and comparable to those in adults. (2) In a process of synaptogenesis neurons grow axons and synapses that contact other neurons, thus generating networks. (3) The initially established networks operate in cooperation and mutual inhibition. The practice of interaction allows control of which networks are appropriate and which are inappropriate. The inappropriate ones must be eliminated, mostly by the *pruning of synapses* and often even of inappropriate or superfluous neurons. Synaptogenesis and synaptic pruning (see Gage and Johnson 2007, pp. 423–425) function in reaction to signals arriving from the environment or from body internal events. (4) In any case these procedures generate internal network structures that sub-specify the innate frame either by adding structure, by modifying it or by eliminating certain parts. They generate the environmental or body-internal *fine-tuning*. (5) It is obvious that these processes lead to increasing complexity of network organization producing global innate structure plus emerging fine-structure. Since the former is determined by the influence of genetic specification and control and the latter by situation-dependent events, one often characterizes the process metaphorically as a *mixture of nature and nurture*.

Another characteristic remains to be discussed. It is the influence of innate *predispositions for growth organization*. The most important genetic predisposition consists in temporal changes of intensity of synaptogenesis and synaptic pruning. The influence that determines specification is strongest at about 1 year until 1 year and a half, with the exception of the pre-frontal cortex, in which synaptogenesis and synaptic pruning, which contribute to differentiation, are particularly strong at about 5 years.

Another very important feature is *myelination* of long-distance axons and fibres. Particularly important is the *myelination of long-distance axons and bundles*. Myelinated axons transmit the signals quicker and more efficiently back and forth between the posterior and the frontal areas of the cortex.

The time stages of development by myelination are represented in Figure 1.7. At birth, the connections to white or grey areas are still very inefficient. Only the dark areas are well established. The transmission of activation potentials over longer cortical distances is slow and not secure since the wrapping over the complete distance by white sheathing cells is not yet completed (see Schwartz 1995, 51–52). The brain organization must wait for the completion of myelination during subsequent years.

This myelination together with the functions that control synaptogenesis and synaptic pruning will contribute to the development of self-organized action with increasing precision, distinct identification and selectivity of action. The completion of these processes takes much time. In a really efficient sense the long-fibre coordination will not function optimally before puberty.[10] As a consequence more complex functions requiring precise selection and attention control cannot be learned before the specific brain connectivities are ripe for them.

The difference between the organization of the innate global brain architecture and the internal principle-based myelinated systems together with usage-based development of the networks of potential and actual cognits show in a typical way how the brain is partially fine-tuned by nurture, i.e., by the environment, but equally partially by nature, due to its enormous basic system of plastic fine-structure at birth and also by innate principles of neural growth processes. To be more precise we should remark that the global architecture attained at birth develops still more efficiency during subsequent years, for instance by myelination of long-distance axons that allow distant fine-tunings to become interactive in efficient ways.

Let me now add a few *remarks about language development and its developmental stages*. In principle, the global brain architecture and the

[10] See J. Fuster (2003) pp. 32–34 and also S.-J. Blakemore and U. Frith (2005) pp. 113–119 and N.M. Gage and M.H. Johnson (2007, 425).

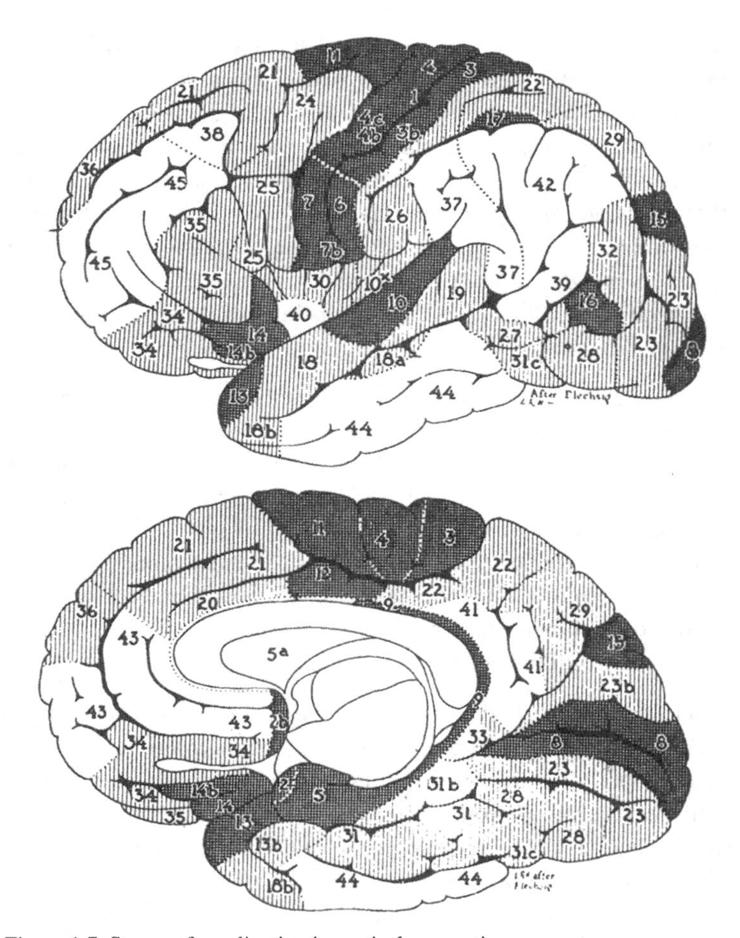

Figure 1.7. Stages of myelination in cortical connections.

development of functional efficiency are determined by the *development of the axon's myelination*, also an important condition of language learning in humans. *The processes of fine-tuning* produce increasingly precise specificity. They are based on the selection of neuronal connections and on the pruning of inappropriate ones. The specific development of connections – thus the development of more complex cognits, that is, pieces of knowledge – are largely determined by typical stages of usage integrating hearing and speaking, learning to read and to write and to think in terms of concrete imaginations and abstract structuring. The specific usage of the systems of cognits leads to different systems of fine-tuned brains in different people. Thus the brains of different speakers and hearers of the same language do not have strictly the same fine-structure. But they are all *very similar* when they speak dialects of the same language, whereas speakers of different languages have a global architecture and long-distance fibres fitting to their language usage, whereas their fine-tunings allow one or the other dialectal and personal specificity. Thus, the acquisition of a particular language and special modifications are realized in infancy, childhood and adolescence by *fine-tuning the primary structures of innate global brain architecture*.

Consequently it will be appropriate to distinguish different levels of fine-structuring:

(1) The *brain's fine-structuring* during the *infant*'s pre-linguistic experience,
(2) The *fine-tuning of early mother-tongue acquisition* in *early childhood*,
(3) The *fine-tuning of knowledge development*, especially the development of more or less advanced *memory-based knowledge frames and their expressions*, and finally
(4) The knowledge frames that are put to use in even more advanced and complex *fine-tunings*, usually organized in the pre-frontal cortex, *allowing* organization and understanding of *narratives, arguments, knowledge and thought* in later childhood, adolescence and adulthood.

Thus the acquisition of *conventional language competence and language-expressed thought relies on different stages of neuronal fine-tuning development*. They properly modify the innate connectivity of basic global structure and the innate determined growth process control. The latter provides at each stage an efficient base for fine-structuring. In this way complex competence like language, knowledge and intelligence result from *sub-structuring of the innate global brain architecture*. To be more precise we should remark that the global architecture attained at birth develops still more efficiency of language and thought during subsequent years, for instance by myelination of long-distance axons that allow distant fine-tunings to become interactive in efficient ways.

The general principles of *neural fine-tuning* have their consequences for language usage. They determine the competence of fluent uttering of language expressions relying on two different *frameworks*: (a) the framework of *automatic*, that is *non-conscious production* of the utterances in the correct form of a given dialect, and (b) the literacy-based conscious production of ideas, plans and knowledge-based systems of thought and argument relying on access, memorizing and restoring of memory data, referring to different knowledge *levels of the worlds*, *self-knowledge* and *autobiographic knowledge and feeling*, *mentalizing of other people's knowledge and feeling*, and of *theoret-*

ical, practical and aesthetic social and religious reasoning. The non-conscious attention-based framework of *basic grammatical language organization* constitutes an efficient system of language expression forms, comprising syntax, morphology and formal semantics as well as the core elements of concrete semantics. The pluri-system of semantics, which necessarily distinguishes the non-conscious and the conscious part, requires more details that will soon be provided. The understanding of core elements plays an important role in *correcting the reductive notion of formalist meaning and semantics* that has become dominant in modern linguistics. In the corrected perspective there are as many frameworks supporting and guiding conscious thought as there are spontaneous archetypes in folk-psychology.

But *literacy-based frameworks* acquire access to many advanced frameworks of understanding often in systematized forms in general descriptions or regimented theories in science and other advanced disciplines. It should be clear that in each literate adult both frameworks are organized non-automatically in the cortical brain. As already explained, automatic language form organization is located at well-known cortical areas, Broca's and Wernicke's, in the left hemisphere. The organization of more advanced competences requires attention-controlled selectivity of structures and requires pre-frontal access to processes of more complex combinations. The *non-conscious processing of semantics* probably relies also on a large set of automatic components, which the pre-frontal cortex may efficiently control whenever necessary. More advanced *conscious semantics* may require a continuous *interaction of pre-frontal cortical areas with practically all other areas of the cortex* and even many nuclei in sub-cortical parts of the brain.

Due to this widespread accessibility in which *language forms are combined with all areas of the brain's organization of concrete semantics*, almost everything that mentally represents the external world in various areas of the brain and also our internal bodily feeling, as well as our practical and theoretical skills and competences, could in principle be expressed in words and sentences. Still we also tend to think that we often can't express things in as much detail as we want to explain. Some brief statements and skeletal word combinations may come to mind and hint at the facts, though without any specific detail. In many cases additional explanations may indicate specifics, when required. But there are knowledge areas that seem to be more limited in accessible details. Here we must be satisfied with global ordinary word meanings like *wonderful*, *infinity*, and hints of "transcendent" knowledge that underlies *aesthetics or thought*. The same holds for *emotional interjections* expressing feelings, for instance in particular situations of illness, of joy or of happiness and so on.

1.7 Models for language in the brain

The previous two sections concentrated on schematic properties of the complete nervous system and of the brain's development. I made some brief remarks about Broca's and Wernicke's areas in section 1.1 and section 1.3. But the status of language organization in the brain certainly needs further interpretation. So it may be worthwhile to recapitulate the development of early models that have been proposed for explaining the functions of Broca's and Wernicke's areas.

Language is often mentioned as one of the functions of the brain, in addition to perception, action, feeling and thought. Figures 1.5 and 1.6 indicate possible areas involved. But a topographic arrangement of areas is not sufficient. We need systematic models of explanation. The first step for interactive modelling was made by Wernicke (1876), some years after brain measurements of physiological and observational aphasia were made by Broca 1864. Some years after Wernicke the neurologist Lichtheim (1885) suggested a model whose basic idea of area interaction is still meaningful.

Pulvermüller presented a detailed commentary of Lichtheim's ideas (Pulvermüller 2002, pp. 35–38). Particularly instructive is the example of word comprehension, here discussed in referring to Figure 1.8. The auditory centre A (Wernicke's area) first receives a signal from the acoustic input. The process of meaningful understanding is instantiated by the activation of a central area B, the conceptual centre. B's activity is caused by signals transmitted via A's connections to B. As soon as the conceptual meaning and an intended reaction are understood signals are transmitted to the centre M, the motor centre that causes the intended articulation. Very interesting is Lichtheim's idea that the same model can be used for explaining the process of thoughtless rep-

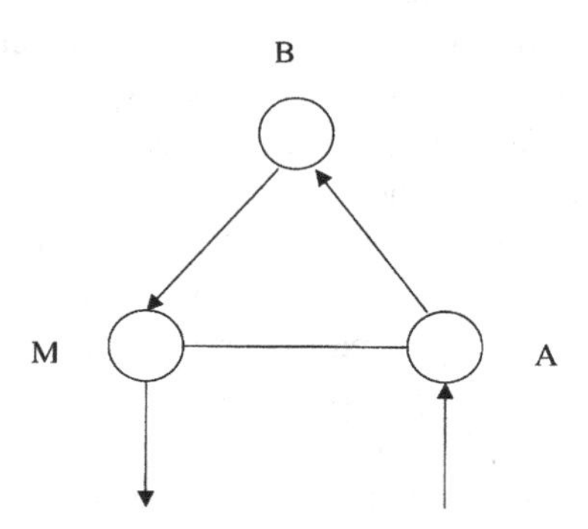

Figure 1.8. Lichtheim's early schema of the interactions of Broca's and Wernicke's areas.

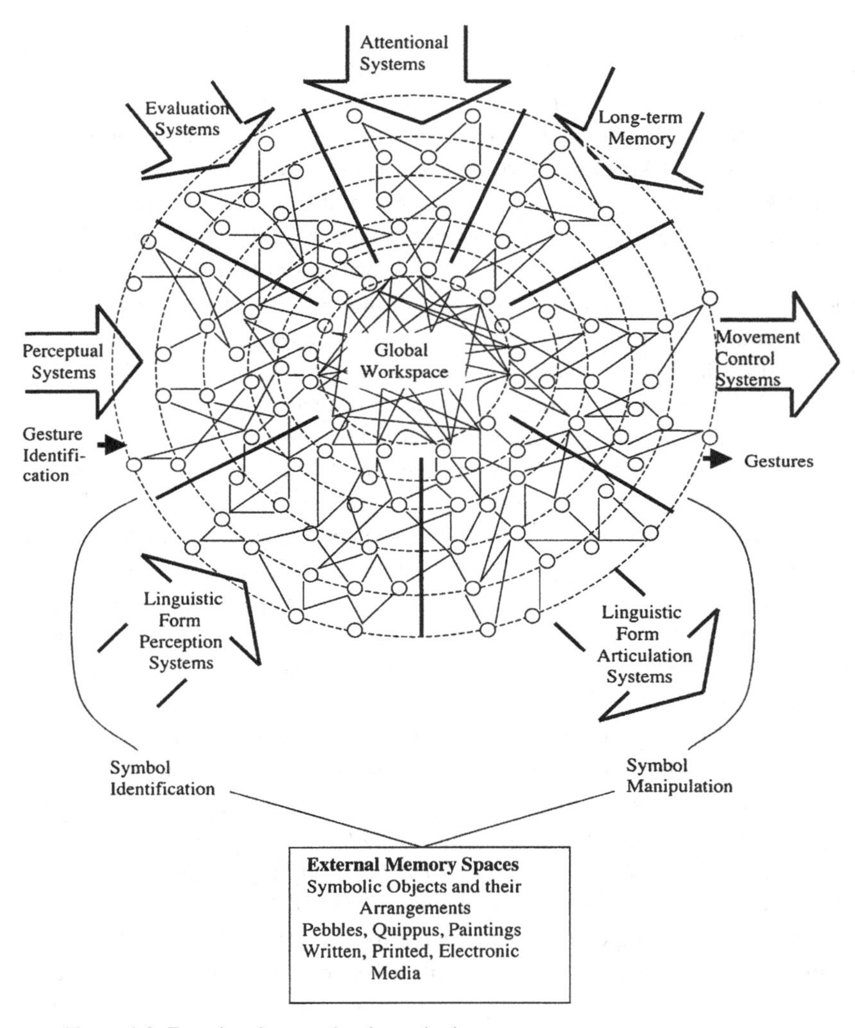

Figure 1.9. Functional network schema in the cortex.

etition of a word. In his interpretation the activity of A, representing the momentary hearing of the word sound can directly cause the activity of centre M without the intermediate activity of B. This is possible due to the direct connection between A and M.

Lichtheim's classic language model and his interpretation was modified by Fuster (2003, pp. 191–192) in an essential points; He argues (ibid.) that a more apt scheme would have the conceptual centre, the assumed substrate of meaning, divided into the two large moieties of association cortex, one in the posterior part of the brain – similar to Wernicke's area – the other in the posterior frontal part – similar to Broca's area. There is one substrate for the perceptual and one for the executive cognit. Moreover Fuster follows a suggestion of Damasio 1994 who considers that the Wernicke's area is more extended than normally assumed. Fuster emphasizes that the more extended Wernicke's area would include associative cortex for stimuli other than those of auditory modality. As a consequence there is a quasi-linguistic form organization level, perhaps contributing to the organization of morphology, and a semantic level that is particularly involved in the *mirror neuron system* that will be discussed in Chapter 3, section 3.3. In any case there is reason to introduce a parallel system of the perception–action systems relevant for concrete meaning and the linguistic form organization system presented in my model of the pair of LM and LF systems in Chapter 2 section 2.4.

The Lichtheim model is global and extremely unspecific concerning B. It comprises the different types of semantics: a) concrete, b) abstract and automatic and c) pre-frontally controlled as well as the principles of their constitution and global interconnections. More modern representations represent the complex character of interacting functional neural networks surrounding a global workspace network. The model of Dehaene, Kerszberg and Changeux (1998) is particularly suggestive. An empirically and theoretically detailed discussion is presented in Dehaene (2009). All points refer to local neural clusters and areas that are widely distributed in the cortex and are connected for allowing interaction. In Figure 1.9 I introduce a *modified version* of this model. As an *essential change I introduce two particular sections marked by linguistic form*. These sections correspond mainly to Broca's and Wernicke's areas serving the core components *of language form organization*: one for structuring of utterance production, the other for sound pattern perception and certain details of grammar and meaning, that is, grammatical structure organization. Processing of *concrete language meaning* will be assumed to occur in many parts of the other sections.[11] Another addition is the indication of influences of gestures that accompany normal speech. I will not discuss them here. The roles

[11] J. Fuster (2003, p. 186) presents a map of cortical areas of the left hemisphere, distinguishing those areas in which lesions often result in severe language disorders, probably due to confusion of language form, from those areas in which lesions result less frequently in language disorders. There is some indication that the latter are concerned with meanings that can, in the case of lesions often be replaced by similar expressions.

of the uppermost area components, corresponding to pre-frontal areas, will be discussed in the next chapter.

At the end of this chapter let me recall again that the patterns of functional architecture and organization are ultimately even better understood when reference to innate structure and postnatal growth is also taken into account.

2

Organizations in complex organisms

2.1 Some philosophers' pan-organic outlook: instead of an introduction

The introduction of the first chapter was a plea for the future development of our functional triangle: linguistics–phenomenological psychology–neuro-cognition. The following illustrations and introductory explanations of brain elements emphasize the enormous challenge of understanding the functional integration of the three disciplines. Serious difficulties confront the inter-disciplinary studies due to the differences of empirical methods and different frameworks for principles of thoughts, theorizing and trying to construct fruitful working models. Since this book is not so much concerned with pre-senting details of measurements and observations the *focus is rather on theo-rizing and clarifying the disciplines' principles* that underlie the conceptual frames for plausible and justified working models.

The differences of interdisciplinary thought are clear. Linguists work with formal constructions considered as plausible and justified working models structuring grammar and systems of meaning rules that describe semantic and pragmatic structure dependencies of lexical words. Their constructions are ar-rangements of conceptual terms in static relational patterns. The set of terms can only contribute to principles of organization, that is to the topic of the present chapter, when *systems of construction rules* are added to the collection of terms, rules whose *operations* identify, combine or separate letter symbols and formal symbol patterns or arrange them in systematically justified patterns representing relations. In modern linguistics these principles of organization have been adapted from organizations introduced in formal logic and formal meta-mathematics during the first half of the past century. Techniques were developed approximately at the same time as computers and computer pro-gramming of digital data processing. Some philosophers understood that these mechanisms were *not really organizations in the sense of living systems*. Thus formalist representations of grammars and lexica cannot be directly parts of a natural organism's organization. Self-critical linguists understood the point but insisted that it is not the task of linguistics to define parts of an organism, not even the parts of the human brain that organize language competence. They think that grammars and lexica define abstract constraints for sound, word and sentence *classifications* and *combination*. The *construction rules* constitute merely formal definitions of the constraint patterns. Their operation has no organizational meaning, not for a computer, and certainly not for parts of the human organism. Linguistic structure analysis has nothing to do with specific processes of neural organizations in the brain.

The brain's neural networks do organize grammatically and lexically correct words and sentences but operate in ways that are quite different from managing symbolic notations of grammatical terms and phonetic features. Should there be some components of the brain that somehow engage in a more appropriate and effective activity? In a functional interpretation certain units in our organism contribute indeed to effective speech behaviour. Section 3 of the first chapter presented the Jakobson–Teuber idea. Feature and category terms could be given "reality" by understanding them as a new kind of *neural module* based on local circuits. The envisaged elementary entity was *not* a single neuron for a feature or a category.

The basic idea should be that organized entities constitute parts of an organism. Compared with symbol combination mechanisms we are confronted with a *radically new idea*: We introduce a descriptive pan-organic system in which each object name denotes, *without any exception*, an organized entity. Whatever can be referred to in terms of this system are organized entities.[1] Could we describe language in terms of this kind of descriptive system? If so there would be organism components that are "living language descriptions," that is elements of a meta-language with dynamic meaning and organization! But this is what we are looking for! We would have a radical contrast to the formal understanding of the essentials of language. Language constraints would obviously be represented by interdependencies or interaction regularities of organized entities – local neural networks, complex neural networks, all of what Fuster would call cognits.

I will present a non-symbolic and non-formal perspective for those who are interested in the fundamental status of organized entity frameworks and their principles. It will be extracted from the small book of a pan-organic philosophy written by a famous philosopher. The following statements present the ideas.

[1] The term pan-organic was invented by N. Rescher (1991).

Finding the name of the philosopher will first be left as a puzzle for the reader, though not without giving a hint: The bracketed numbers refer to the locations in the author's small book. Readers who don't like puzzles of this type may go directly to the end of this section where the puzzle's solution will be presented. Here is the text:

(1) Each created unit is subject to *change,* and this change is continuous in each one.
(2) Each *organic* body is a *machine of nature* (M64). This organic "machine" of nature is *infinitely sub-divided into parts of parts* according to biological body constitution. *Each part is itself a biological organic unit having an operative function of its own* (M65–M69).
(3) Even in the last piece of matter there is a *whole world of dynamic sub-creatures* (M66).
(4) Each complete body has a *dynamic functional design principle (a mental* functional substance, being the design principle of the complete machine of nature (M64)), characterizing the body's constitution and its temporal existence (M70, PNG 1).
(5) A simpler "incorporated automaton", like most sub-division units (cells, cell complex units etc.), should still be understood as an organism (with its proper dynamic functional design principle). But, though organized entities, these simpler units would not be called animate beings or animals (M63).
(6) The sub-division unit has itself a dynamic functional principle that has, however, only the status of *relative* functional dominance; it is called relative dominance, since it determines only a sub-sub-division whose processing must correlate with the basic functional principle of the complete body (M70).
(7) It is important to understand that the whole system of sub-division is a hierarchy of sub-units. In the hierarchy of parts and functional dependency, all design principles of unit and sub-units are in "harmony" relative to the "laws" that characterize the central design principle (M78). That means that, though the dominant functional unit acts according to its own "mental laws," the body is determined by the collective system of the relative "biological laws" of the parts composing the body's hierarchy constitution. The correspondence of the dynamic function principles of the units builds the "mind–body" harmony (M78). The mental centre unit together with its system of parts appears as an infinitely detailed "divine artifice" (M64).
(8) The coherent activity of a single body relies on the *biological linkage of the activities of the simpler sub-units*, belonging to a single animal's

functional unit. This is similar to how the collective system of animals in a group relies on the interlinkage of the animals' dominant dynamic function principles (i.e., souls) to form the social collection (M56, M62).

(9) The previous notion of interlinkage corresponds to intercommunication (PNG3) and suggests the notion of *interconnection* on the animal's subdivision parts that build biological groups (PNG3, M61). Bodily parts affect each other by touch – directly or indirectly (M61). This "touch" of parts is normally due to their biological relation constituted by contact-sub-parts (like the "organs" or the spreading out limbs in animals, which we would now rather exemplify by axons, synapses and dendrites spreading out from nerve cells). Such an interconnection system constitutes a network "intercommunication" that extends to any distance, however great (M61).

(10) The effects are interdependent *changes of states* of the interconnected units. All bodies are in continuous change, because at each moment functional sub-units, sub-sub-units etc. are in perpetual flux, some entering and others leaving the body (M71, M72).

(11) *Up until now we* concentrated on body-based constituency and organization in which the *dynamic functional design principles* and their interdependencies *play a fundamental role*. The integration and interdependency of many dynamic change principles (down to infinitesimal elements) is biological. It is not a unified law system as is usual in typical physical systems. Thus the perspective is biological and not physical or mechanistic (M64). We should now concentrate on the status of a *dynamic functional design principle:*

(12) Here is a recent philosopher's commentary to M11: our philosopher considers the *internal descriptive design principles* as dynamic. He postulates that they are akin to the *algebraic rule* that generates a numerical series or (better yet) to the generating equation for a continuous curve. This systemic unfolding of its own successive states – the whole history of its particular actions, so to speak – serves to endow each biological unit with its own particular individuality. Here is a philosopher of process, who sets himself apart from the succession of theorists who see time and change as insignificant and somehow illusory features of a fundamental timeless and unchanging universe.

(13) In an algebraic model changing units are numbers. What are the descriptively basic changes in the dynamics of biological units? Fundamentally there is the *dynamic principle* of change (*entelechy, Force*). There are inner biological unit state representations (*perception representations, cognits*) and biological unit force's execution-representations expressing an inner change of state (action as force execution) (M11).

(14) The more abstract notion of inner state or inner change of state has the following more concrete interpretation: The inner states of a biological unit *represent* at each moment the *status of the biological unit's external world* and the *biological unit's changes* contributed through its relation to the world.

(15) Beyond the principle of representation and of representation change there must also be an *internal complexity state of the biological unit* that would produce, so to speak, the specification and the functional representation *variety of the system of the complete biological unit* (M12).

(16) The transitory states that *enfold* and represent a *multiplicity of complexity in a biological unit* instantiate the *general kind of representation*. More specifically it would be necessary to distinguish in animals, and in particular in humans, *conscious representations from non-conscious representations* (M14).

(17) We ourselves – as a specific kind of animal – experience multiplicity of perception representations when we find that the slightest thought of which we are *conscious* in ourselves *enfolds a variety of features in the perceived object* (M16). This may also be applied to non-conscious "perception" representation. In any case we can say that, insofar as a biological unit's description determines all respects of object characterization, we may say that this functional description is *descriptively dominant.* In this sense of descriptive dominance of representation competence we are justified to call it *mental.*

(18) If the dynamic functional design principle of a complete bodily unit involves sentience of the body, the complete body is an animal (M19).

(19) Nature has given *heightened perception representation* to animals by the care she has taken to furnish them with organs, which collect many rays of light or many vibrations of air, and process them further to make them more effective through their unification and *internal processing*. There is something similar in smell, taste and touch, and perhaps in many other senses (M25).

(20) Moreover, the next stage of complexity is given by *memorizing experiences*. Memory provides a kind of connectedness that resembles reason, for example, experiences leading to habitudes and skills (M27) or situation impressions in animals. We are all mere empirics in three quarters of our actions (M28, M20).

(21) But the knowledge of necessary and internal truths is what distinguishes us from mere animals and provides us with *reason* and the sciences, and elevates us also to the knowledge of ourselves. This dynamic organization may be called the rational soul or spirit (M29, M30).

These were statements of a philosopher who lived 300 years ago. You may have already guessed that our philosopher is *G.W. Leibniz* and that the little book is his *Monadology*.[2] A *coherent summary* may be helpful: An organic body is analysed as a machine of nature. Since it is not merely a constructed machine according to human design, it is infinitely sub-divided into parts, parts of parts etc., down to infinitesimally small units. (Note that at the same time as Newton, our philosopher invented the infinitesimal calculus.) Each unit – the complete organic body and each of its macro-, micro- or infinitesimal parts – has a descriptive function characterizing functional mental design. The constitution of the complete organic body and of the collection of the component designs form, together with their biological linkages and interconnections, a hierarchy of system parts in functional harmony. The mutual linkages and interconnections determine the complete units' internal states accounting for external situations as well as for changes of internal constitution and external world structure. Functionally the dynamic biological linkages instantiate the internal forces of organization execution as well as of change and their dependency on dynamic knowledge accounts of the external situation. In their principle the dynamic organizations and functions of the units exist without end, though there is continuous biological change of the complete units' constituent parts. Different *stages of complexity* of animal component units should be distinguished: (a) Simple reactivity units (like cells and simple cell complexes), (b) units with memorized skills based on automatized organization of perception and action, (c) units showing body-based sentience, and (d) units having moreover the competence of thought, reasoning and planning (only existing in rational animals, such as humans).

Note that our philosopher was not an opponent of formalisms. On the contrary he was also the inventor of precise elements of formal logic, based on formal symbol combinations. The ultimate development during the first half of the past century contributed basically to the development of theoretical linguistics during the second half. He thus contributed to the origins of modern logic and modern theoretical linguistics.

Indeed linguists often still insist on the idea that the abstract representations define the essentials of language, and many theoretical linguists see no need to find out how the psychological mind and the biological brain organize language. This is a statement that obviously separates this discipline from our functional triangle and thus attacks the basic aims of the present book.

[2] More details are found in N. Rescher (1991). The previous list of statements contains also a few references to Leibniz' parallel book *Principles of Nature and Grace*, here abbreviated by PNG.

Which is the appropriate defence? Here is my proposal in two steps. The first is most simple: I merely accept formal or other schematic descriptions used by linguists, when they are understood as classificatory and schematic relation-defining representations that focus the laws or rules of the complete system of a language, perhaps even fundamental principles of defining the characteristics of natural languages in general.

What is the next step that would combine the language's structure representations in the complete functional triangle with the two other disciplines? I require that the solution should not merely construct a more comprehensive integrative formalism but instead a philosophically correct explanation of the mutual interdependence of the three disciplines. I basically apply an explanation that is similar to Kant's (1790) explanation on the three levels. On the first level we consider the situations, events and things by which we are affected, bodily external or internal, and which produce in us immediate perceptions and feelings that are in some cases recalled from memory. The whole range of external data together with our internal perception or perception memory is the *range of sensory data affection.*[3]

The next level is concrete "reason". It is our mental power to produce conceptual images that provide ordered complexes of our given or constructed data affection. Our common sense "reason" constitutes, together with the power of imaging, concrete common sense knowledge. Varieties of these also occur in animals. In addition to this more or less immediate "lower" form of knowledge there are in principle also varieties of our "upper" form of knowledge. Deviating from Leibniz' view I accept that *internal language knowledge* is a particularly important component of our "upper" form of knowledge, that is, a *mind internal fact* rather than spoken or written external tools orienting the mind. Based on language knowledge and on thought, even higher forms of this "upper" form of knowledge are constructed by our "upper" varieties of reason, for instance, those that lead to our scientific understanding of the *laws of the world's* reality. In their precise variety they form the *scientific knowledge of advanced reason*. Applying this knowledge to the perceptual facts of the world they become elements of a systematically ordered world in our psychological acts of understanding.

Shouldn't we apply this model also to language and give knowledge of language, with its set of grammatical rules, a similar status to the laws of physics? The linguistic rule system – or, as we might also say: the linguistic reason – determines the ordered "world" of linguistic facts, whenever we apply

[3] Kant's term is "Anschauung."

this "reason" in a regular way to language data. The role of laws in physical theory correspond to rules in grammatical and lexical theory in linguistics. In both cases our combinations of reason and given data are presented on two levels: *mental thought* and mind internally received and differentiated *perceptual data* on the one hand, and law-based or *rule-based theory* on the other.

Our next step is the most important one in the present context. It transcends the limited aspect of normal empirically based formal theory. It opens perspectives of thought that bridge the gap from the "world" structured by formal laws or rules of the physical world to the "world" that was presented in Leibniz' pan-organic outlook. Here also my arguments rely primarily on Kant. In his considerations about transcendental philosophical foundations he adds to the three empirical science-oriented levels (1) the *sensory data identification system* (*Anschauung*), (2) *reason* (*Verstand*) and (3) the general base of *pure science-directed philosophical foundation* (*Vernunft*). Reason concentrates on our knowledge of *laws* or of established *rules* that are systematically integrated by philosophical foundation. In addition to both and between them Kant requires thinking in view of philosophical foundation of judgement. In this perspective he considers, in addition to aesthetics, the principle of teleological appropriateness in our understanding of biological nature (teleological *Urteilskraft*).[4] Our thoughts tell us that the appropriate entities in biological nature are those that are *organized components*. But also the complete organism and in particular a reasoning person's body is an organized entity, that has its own end in its self. For my interdisciplinary functional triple it is important to understand the *bridge from laws and formal rules* to units of *activity in a biological body and the mind of both own-self and other-self experience*.

2.2 The neuroscientist's basic reflections

Some modern approaches for unifying cognition and feeling experiences are in many respects similar to our philosopher's framework. The neuroscientist Damasio's last book (Damasio 2003, pp. 194, 210–215) related his principled background to Spinoza's philosophy. Though Spinoza and Leibniz differed in their principled views about God, their "organic outlooks" at things, individuals and dynamic world structure are relatively similar. The approach of the neuroscientist Fuster (1790) is also rather open-minded in considering mental phenomena, in particular the proposals of gestalt theoretical psychology. In

[4] I. Kant, *Critique of Teleological Judgement*, section 65.

both cases dynamic interdisciplinary functionality of mind and brain aim at a common framework.

In fact the models of the neuroscientists have recently added systematic sub-structuring to the 300-year-old idea of the organism's constitution and its structuring into infinite sub-divisions, sub-divisions of sub-divisions etc. down to micro-activity units. Let me first enumerate the scales of their organic stage levels. According to our *functional principle the phenomenological or mental aspect will be given the functionally primary status* in the a.-lines of the following list. The corresponding *neural network aspects* are presented in the b.-lines. Note that following Fuster (2002, p. 226) the a.-lines, from 2.a downward, are types of pieces *of knowledge*, or of *cognits*. He emphasizes that, from a point of neurobiology, knowledge, memory and perception–action organization, attention, language and intelligence share the same neural substrate: an immense array of cortical networks or cognits that contain in their *structural mesh* the informal content of all functions. The corresponding neurocognitive networks are listed in the b.-lines, from 2.b. downward.

1. a. the complete mental system of human cognitive competences,
   b. (human cortex + limbic annexes),
2. a. mental representations of perception, memory, attention, language intelligence,
   b. (interactions of areas),
3. a. mental representations of complexes (perceptual situations, events and processes),
   b. (structure combination networks),
4. a. mental representations of complex feature or category arrangements,
   b. (complex module networks),
5. a. mental representation of single feature or category,
   b. (local neural cluster networks),
6. a. elementary function sub-units,
   b. (neurons),
7. a. elementary sub-sub-units,
   b. (nerve components: soma, synapses, axons and dendrites).

We certainly could proceed further down to molecular biology, or perhaps even to the continuum description of infinitesimals, an idea that would please Leibniz.

The previous list tried to show that the biological methodology and the mental studies are nearer to each other in the sense that both, though different in methodology, rely on the dynamical description of the activity units. The process changes are not determined by globally applied laws but are instead based on "biological forces" and "internal states" organized by internal interactions of the units and neural complexes of the body. Their biological dynamism corresponds to mental drive, whether conscious or basically nonconscious. This aspect indicates how to extend the specific mutual bridge between phenomenological mind and neurocognitive brain to the third partner, that is the mental structure specifications of language form, as far as their *category configurations* also represent pieces of knowledge and thus give guidelines for bridging, perhaps more or less following and extending the basic principle of Jakobson and Teuber. We should look for linguistic knowledge pieces (cognits) represented as an interaction system of dynamic units of the knowledge application dynamic. A unit is in the dynamic state "waiting" for possible action. An appropriate contingency moment activates the neural *energeia* – the momentary realized activity. When several cognitive network components remain simultaneously or repetitively in their energeia state for a short time we will say that they organize their *mutual binding*, thus forming a cognit complex.

Following Fuster's account we concentrated on *cognits* and *neurocognitive networks*. This perspective must be further extended. Neurons and neuronal connections are not only distributed in the cortex. There are many other parts that interact with the cortical centre. Let me briefly repeat what was already set in Chapter 1. The global characteristics theory states that the nervous system has two components: The central nervous system (forebrain, with the cerebral cortex, midbrain, hindbrain and the spinal cord), and the peripheral nervous system (somatic division and autonomic division) which is widely distributed internal to the body. The latter divisions organize the body's states and influence self-feeling and emotions, that is in clear contrast to cognition. But in principle the interactions of smaller and larger configurations of nerve-cell networks will also constitute a hierarchy of neural sub-divisions, sub-divisions of sub-divisions etc. The important difference in the mental aspect is that we must distinguish the "*external world-oriented*" perception–action system and the *body-based self-experienced self-presence feeling*.

Concerning the cortex external world we should distinguish the externally oriented and the internally oriented phenomena:

World units (as typically described in objective description, either in scientific or in phenomenological terms: structures, events, processes, locations, times etc.),

Relation of functional *elements of brain design* to *the facts of the world,*

Relation of functional brain design elements (cortical and cognitive knowledge representation units) to *body internal facts* (visceral and emotional) (Damasio 1999, p. 40–41; Fuster 2003, p. 137–139).

2.3 Fuster's perception–action cycle: a basic format for studying brain architecture

So far we have emphasized the interactive dynamics of mental pieces of knowledge and their functional counterpart in neurocognitive networks connecting distributions of different local neural clusters. In a first reflection, the cortex organization does not appear to be much less confused than the neural clusters represented in Figure 1.1. More precise measurements, analyses and systematizing working models demonstrated however that there are in fact some principles that determine the comprehensive organization: a hierarchy of functional levels ordered from periphery to near; from different modality-specific feature organization (visual, auditory and haptic) in the *phyletic* cortex, to the processes of the *association cortex* and, finally, to the level of more general *polymodal* combination – for instance, of the perception activity of the word sound *tree* with the locally different activity of identifying a tree's visual perception. On top of these levels of usually automatic processing complexes there are the pre-frontal areas organizing selective control, as already mentioned in the previous chapter.

This was a brief account of higher organization frames in the cortical perception–*action* system as presented by Fuster (2003). Damasio's (1999) layers are similar but his terminology is more mentally expressed when he explains the system with reference to *acts of recall*: The top-most *pre-frontal level* provides memory specifying symbol terms or symbol complexes that activate on the lower layers' automatic polymodal and associative processes, which in pre-frontal activation contribute to the conscious access of a memory image.

In Fuster's levels and in Damasio's layers two parallel hierarchies are interdependent by mutual interaction, the strictly perception-organizing bottom-up arrangement and the action executive top-down organizing. Figure 2.1 presents Fuster's schema.

In its details Fuster's basic model (2004) concentrates on the organization of world perception but also emphasizes the parallel connection. It shows how *perception is necessarily connected with action* in situations of the world model of a perception–action cycle. Microelectrode research by Rizzolati (2004) and his group has demonstrated the empirical foundation of the systematic posterior–frontal interdependency. The empirical facts and explanations will be discussed in Chapter 3, section 3. In the general perspective, the basic cortical organization is constantly updating its motor systems by means of sensory input, and telling reciprocally the sensory system, by way of motor signals indicating what to expect in the next perception.

Thus functionally the nervous system is always cycling information between perception input and action output channels, to keep the sensory and

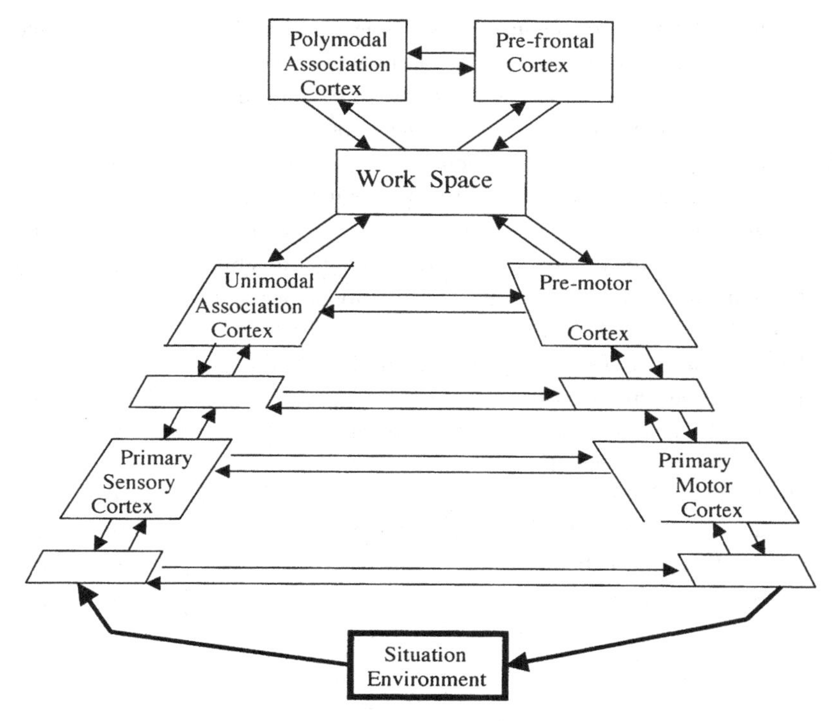

Figure 2.1. The system of perception–action cycle according to Fuster.

motor world in synchrony. It seems that in general there is *no perception without action experience and no action experience without at least a vague perception image*. Figure 1.5 in Chapter 1 indicated already the growth of the rich system of fasciculi in the infant's cortex. The fully developed system connects perception areas in the posterior cortex with action execution operating in the frontal cortex.

Let me summarize the function of this information flow active in the perception–action hierarchy. Lower levels in the sensory hierarchy represent more directly perceptual feature cognits. On the same lower level the motor counterparts organize movement-specific muscle activations reacting to and controlled by situation structure differentiated perception features and categories. On higher levels the hierarchy processing produces more and more combinational or association-determined cognits that represent and signal general or abstracted complex determining features or categories.

In connection with the multimodal level in the middle I mentioned the concrete semantic relation of sound and vision. It applies in particular to linguistics.

The sound of *tree* relates mentally to the object tree, and correspondingly the cortical perception area of sound is in an interactive relation with the perception area of a tree figure. As soon as the brain's operation establishes the distant correspondence of the sound-meaning connection it synchronizes and *stabilizes* this relation *for a short time*, thus generating Saussure's word unit of "significant" and "signifié" not in written symbol combination but in temporal connectivity synchronization. In neurocognition one calls this stabilization the *binding of the cortical cluster's activity*. It is interesting that this binding does not only relate momentary perceptions but also the sound perception with activated memory of the denoted object.

This semantic binding was very well measured and illustrated by Pulvermüller (2002). It will be sufficient to present and discuss the global schemata of Figure 2.2. The schemata represent Pulvermüller's MEG (magnetoencephalographic) measurements of word perception.

The small circles represent local neuron clusters and represent a collection of linguistic category cognits. Thus Figure 2.2.b represents those circle complexes that schematically indicate locations of MEG results when only

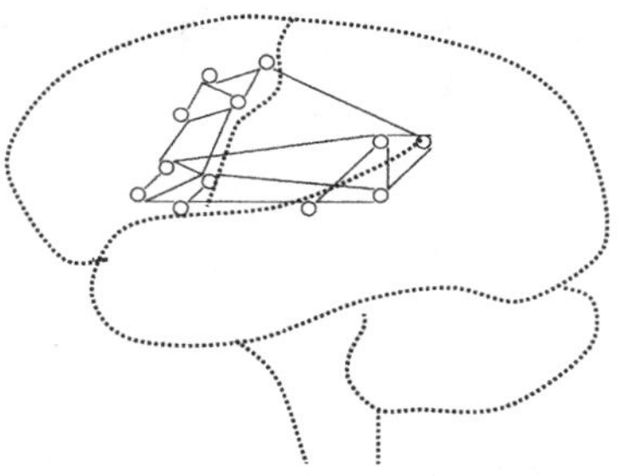

a. Activations of movement words' form and meaning organization

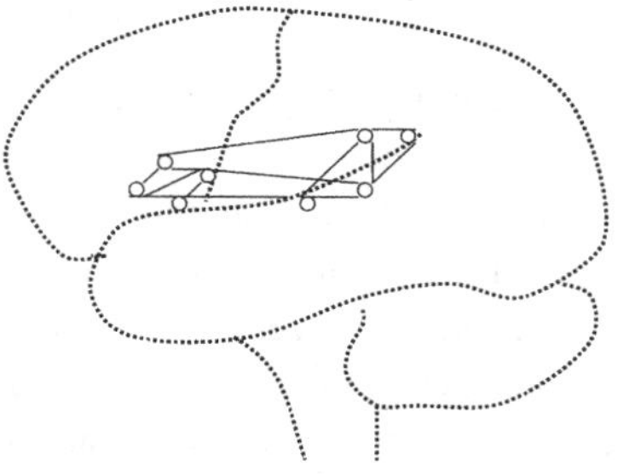

b. Activations of word form organization areas

Figure 2.2. Some semantic binding connections.

sound patterns of words are perceived. Figure 2.2.a shows, in addition to the activity presented in Figure 2.2.b, an activity in a higher part of the frontal area. This area is typical for organizations of leg movements, such as *walk*, *kick* etc.

After having learned the synchronized binding of separated and distant neural clusters representing words' sound–meaning correspondence we are prepared to return to the general systematic of the perception–action cycle. Its system provides the *spontaneous non-conscious organization* of *common sense process complexes,* normal sound meaning relation, perception of simultaneous situations as well as own action engagement. All of them organize "pieces of knowledge" (cognits) as in characteristic everyday use of common sense. The generally correct arrangement of words may be produced without consciousness. A rough idea may have been selected as a symbol in the pre-frontal cortex and signalled to the automatic parts of language form organization, namely Broca's and Wernicke's areas. They will do their job without requiring further conscious control. The common sense competence works spontaneously and is automatic. It is important to understand that this self-organization of the perception–action cycle is characteristic for the complex processing per se, excepting the pre-frontal cortex organizing attention and selection.

In the whole functional framework, bottom-up and top-down, language form action is related to language form perception and perception to action. This should be similar to Fuster's concrete perception–action schema of Figure 2.1, functionally explaining the cortical connection discussed with reference to Figure 1.7. An organization schema for language form should be similar to the one of Figure 2.1. Here also all connections are reciprocal such that, in the complete system, information flows both ways and mostly generates cycles. The consequence is that we have now two parallel ladders, one for the form the other for the concrete meaning, both interacting in meaningful language utterance organization. Figure 2.3 provides the schema.

2.4 LF-cognits and M-cognits in the perception–action framework

This proposal is relatively simple in a direct application of the *layer principle to linguistic form organization.* There is some plausibility in this idea. But on the other hand, its explanatory status may not be as simple as the principle makes believe. After all, the non-human primates, our neighbour species in development, do not acquire language or structurally organized forms of dance or of music. There are some species of birds that learn differentiated birds' call, but it is difficult to discover sufficient similarity. There must be some *innate pre-condition of our human perception–action cycle* that makes it to acquire a language. Even the general organization principle must be the same as the perception–action organization of non-linguistic phenomena as presented in Figure 2.1, but on the other hand something specific must be involved in

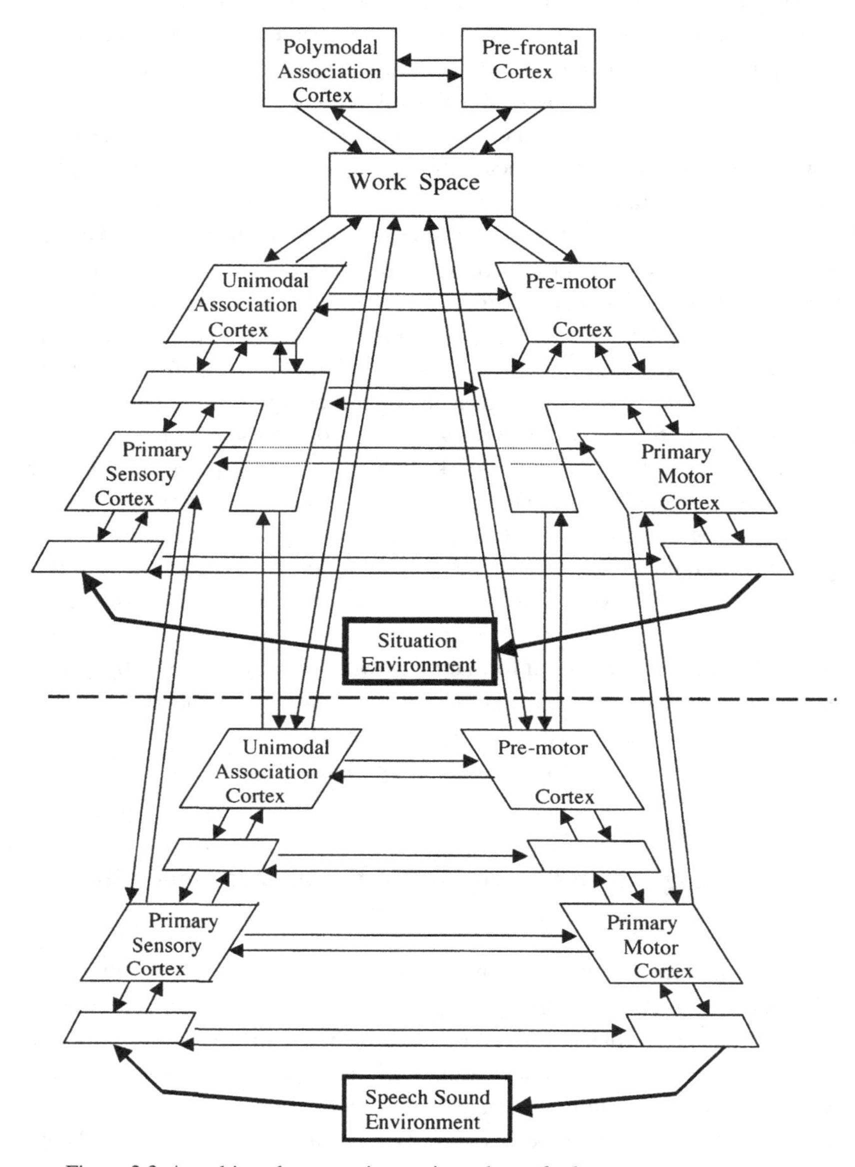

Figure 2.3. A multi-cycle perception–action schema for language.

order to justify representing our human situation by a *double perception–action ladder* like the one in Figure 2.3

The two ladders are four levels of hierarchies integrated by parallel frontal–posterior connections. Two language-assigned hierarchies contain the language form organizer areas, the first for perception such that perception layers comprise layers of sound pattern processing, lexical item processing and perhaps syntactic processing in the central part of Wernicke's area. The second is an executive hierarchy. It contains structured articulation production in Broca's area on some associative ladder level. From the linguistic point of view this *"hierarchy of form"* for sound perception and sound articulation needs another "ladder", the hierarchy of meanings. Here an additional distinction is necessary. First there are the lexical meanings specified by meaning relations as in a good dictionary. We must assume that many of these are similar, and perhaps near to the language form organization network. But, if the situation in which the meaning of a word or sentence is determined by the concrete meaning aspects of the situation, that is, the specific objects, persons, events, actions and so on, then the meanings are structured, characterized and identified by the momentarily relevant components of the normal perception–action system that also operates in any concrete meaning. We therefore must conclude that the organization of concrete semantics is one of the tasks of the perception–action cycle. The representation of the appropriate *concrete meaning is activated in the perception–action cortex.*

I think we are indeed justified to introduce the new ladder parallel, the language form organization system, as a specific partner to Fuster's ladder. The new complex has its own status, and its components should be given their characteristic names. We may say that, since it determines simple or complex linguistic form elements, it should be called the system of LF-*cognits* (linguistic form cognits) related to the corresponding LF-*neurocognition network*. The latter has a dynamic character as, in general, do the perception–action networks. A non-activated LF-network represents a *potential* cognit activity. When connection-influencing conditions activate it, its activation realizes the presence of its cognit. Linguistic form patterns from the lower levels containing the simplest phonetic and phonological features lead to the next level of syntactic, morphological and formal semantic features. All of these complexes are LF-cognits. Thus the functional correspondence of the biological dynamism in the network to the mental characteristics of the cognits is established.

We must now account for the fact that the distribution of cortical and subcortical organization of language form is completely different from that required for *concrete meanings*. Meaning organization is similar to concrete knowledge involving processes of perception, action, imagination, focusing attention, self-feeling, other mind imagination, rules of social knowledge and behaviour. Their concrete experience is not determined by formal meaning-relations but is rather organized by networks of the normal perception–action cycle. It is in this system that meaningful facts are determined in many cortical areas that organize concrete external and internal behaviour and experience. Contrary to lexical meanings, these meanings are lexical meaning relations that can be determined in the structure-organizing LF-system. I think that M-cognit is an appropriate term for any concrete meaning. Consequently, semantics has separate locations: M-cognits that are concrete meanings and meaning relations that are organized in parts of the LF-system.

Concrete meaning or M-cognits have indeed a special character. They have their cortical and sub-cortical counterparts, as well as all situation contexts. They also require much more complex neural network connectivity than language form organization. The neural processes organizing the neurocognitive networks for M-cognits, namely *perceptions, actions, imaginations, feelings, planning acts, recalling people* etc., are distributed over the cortex and parts of sub-cortical areas. Considering these aspects it is quite clear that this neurocognitive separation of the LF part from the concrete M-part is of fundamental relevance for language in the brain: *Language form organization* (LF-cognits) is located *in limited areas of the cortex for concrete meaning;* therefore, M-cognits may be activated in almost any part of the nervous system and not only in the cortex. *Emotion and feeling* may be elements of meaning and thus are *organized partially by the body-distributed nervous system.* There is practically no component of the nervous system that does not "wait" for signals of a relevant context for its activation. Thus contexts in the wide sense are the basic element for selecting and activating the proper meanings, feelings or thoughts. The enormously distributed association system of M-cognits is more complex than that of LF-cognits. As a consequence lesions in the LF-areas are more easily disturbed than lesions in the M-areas. The reason is that disturbed M-cognits may be substituted or re-organized by the activities of others.

2.5 Perception–action cycle as a base of memory

The dictionary says: If you know a language you have learned it and can speak it. Do you have it in your memory? The dictionary says, "Your memory is your ability to retain and recall information, ideas, images and thoughts." But this is probably not what you do or did while learning your language. In a sense you may have done just this in later childhood while further developing your earlier knowledge of language. This was learned however in practical use of whatever is accessible in concrete situations. In this stage you did not keep your words or sentences in your memory, but rather you fixed what they *meant*. Learning language is now learning *to use* it. And this is similar to learning other skills such as walking, bicycling etc. Memory in the ordinary sense of storing and explicitly recalling data is not involved.

For adults a common sense idea is somehow that memory is what could also be written *down in notes.* For the educated adult *memory in mind appears to be like memory on paper.* In this view language, namely in written form, indeed generates reliable memory data. Even philosophers, speaking of original knowledge as *tabula rasa,* seem to have thought that the adult's memory is a store for imprinted results of experiences. Modern computer designers followed this idea in constructing programme-accessible digital data stores, so-called computer memories. In computers appropriate processing and even reasoning rely on stored formal symbol configurations coded and accessible in bit patterns.

Cognitive science and neuropsychology, however, have now made it imperative to revise the view of programmed processing of stored data patterns and codes of data arrangements. The dynamic aspect is completely concentrated in a central processor unit or some other variety of Turing's original idea of a mechanism reading, writing and modifying sequences of formal symbols on a storage tape. In my view it is of fundamental importance to understand the principled difference between Turing mechanisms and the appropriate view of neurocognitive science. The understanding of the necessary revision is particularly difficult for linguists who are used to thinking in written data representations and texts. Since the time of the phonetic alphabet's invention even the serious understanding of sound events relies on written data. Thinking in terms of dynamic action networks generating and de-activating activity states and depending on given activity contexts is rather strange to the linguist. Still thinking in terms of activity is in fact possible in studying speech acts and acts of communications. But in this perspective the acts are guided by intention and selective attentions. A transfer from "communication" of neural units to reactive interactions of neural clusters lacks the reference to the linguist's as well as the logicians' mental structure understanding. Searle (1983 p. 269–270) has already indicated that intentional actions and neural causality should be understood as correlated.

The *necessary revision* may best be understood by considering the relation of memory and knowledge as understood by Fuster and explained in his cortical working model. As already mentioned the core idea is that pieces of knowledge are named cognits and represented by neurocognitive networks. In order to approach the correct meaning of this proposal the following clarifications are necessary: even in its mental view, there is a basic difference between pieces of *knowing-that* from pieces of *knowing-how.* Being able to balance the bicycle in the course of cycling is a piece of know-how. It may become active in given contingencies. Note that in contrast to knowing-that, this activation does *not require an act of conscious recall* but merely a situation appropriate application of a piece of spontaneous competence. In other words, each piece of this competence is a possibility of action—a δυναμισ—and its automatic activation is a momentary realization of this "waiting" possibility—in a ενεργεια activity in terms of the Greek philosophy. In contrast to the dynamic unit a piece of *knowing-that* is *not a possibility of action* but at best a possibility of being an object of recall, of being expressed in an accessing act of thought or imagination, or, explicitly, in uttering externally words or sentences. It is in this way that we must *distinguish* between *dynamic cognits* and *pieces of static configurations representing factual knowledge.*

Let me conclude: Knowing-how to do something is to rely on a network of dynamic cognits and how each dynamic cognit complex is activated or "energized" in an appropriate condition or intention. Normally, it becomes active given that the appropriate conditions are themselves also specific network activities that are connected to the knowledge network. In this way a mental cognit corresponds to a realization in some biological cognitive network unit or a network complex in a field of situation cognits. If this is accepted it is a principled necessity that, in the spontaneous or automatic aspects of our brain's language competence, linguistic features or categories, feature- or category-combinations or associations, and even category structure complexes are realized in situation-conditioned cognitive networks, and so is the brain's organization of our conscious word- or sentence-sound images we just heard or recalled. The problem is that we do not yet know the details of their growth, ultimate network connections and hormonal support that would allow us to define a mathematical theory of their dynamic.

We are now prepared to understand Fuster's statement that memory is made up of mental cognits and their biological correspondence consisting of neural networks. These memory networks operate dynamically and generate at appropriate moments cognit activity supported by other cognit's cognitive networks that represent background cognits representing concrete situations. They may also answer specific signals of pre-frontal memory by their own activity. This kind of activation is often *misleadingly called retrieval* in thought processes, though there is no agent in the brain that "retrieves" something and that would "look" for a "readable plan" to be executed. Fuster's careful analyses can demonstrate that neurobiology, knowledge, memory and perception share the same neural substrate: an immense array of cortical networks that contain in their structural mesh the informational content of all three functions. Thus Fuster's principle is justified by stating that knowledge is the power of dynamic memory and structured dynamic network power is knowledge.

2.6 Large scale models of functional neuroanatomies

So far we have concentrated mainly on the question of how neural clusters or neural complexes could be understood as dynamic bases that dynamically generate energized momentary activity patterns that fit the activity of other context patterns which represent the situation. Following Fuster we characterized systematic arrangements of dynamic interaction in a hierarchy of a normal perception–action cycle parallel to a hierarchy of a language form perception–language form production cycle. The second hierarchy comprises Broca's and Wernicke's areas. The reference to Broca's and Wernicke's areas gives rise to the question of how the dynamic processes are distributed in the neuroanatomy of the cortex. Given the current limitations of brain imaging's local resolution rather invites the development of models that refer to the more global complexes of the cortical brain anatomy. This was indeed considered in models of functional neuroanatomy developed by Hickok and Poeppel (2007) as well as by Ben Shalom and Poeppel (2008). The first article discusses a speech-processing model in which the roles of different connection activity streams in the cortex in the left and right hemisphere are characterized. In this model there are, in principle, two activity streams in each hemisphere: There is a lower, namely ventral, stream that processes speech signals for comprehension, which shows equal degrees of activity but determines different roles in the left and right hemisphere streams. The higher, namely dorsal, cortical streams are concerned with analysing concrete tasks and conditions that are focused differently in the two hemispheres. It is clear that the roles and tasks of processes in the left hemisphere are dominant for speech.

2.7 Pre-frontal attention access to the perception–action memory

So far we have mainly considered implicit and automatically organized memory. We now are going to consider explicit memory. We shall see that, from the perception–action point of view, there is not really a difference of content that occurs in the core system. The difference is only that the activation occurs either in implicit (non-conscious) or in explicit (conscious) memory. In the explicit activated memory, specific usage is organized by attention- and consciousness-oriented selectivity and consideration of the dynamic of thought. The essential point in this case is that contents can also be generated by implicit memory and are now selected and sometimes consciously combined or synthesized, as Ben Shalom and Poeppel say. Elements of memory's normal self-organizing process now become influenced by selectivity of the pre-frontal cortex. Self-organizing is for instance the normal case for the *automatic organization of grammatically correct processing* or understanding of fluent speech in a mother tongue. That is, the organization of language form usually also involves understanding its concrete common sense meaning.

Which are the typical situations of *selective* influence from the pre-frontal cortex? Consider an idea that you want to express and communicate in some correct sentences. The idea may just be present as a skeletal collection of idea-related, intention-symbolizing core words. Consider the various explanations of Fuster (2003, pp. 96, 106) and Langacker's (2008) notion of *skeletal word collections* as bases of *grammatical "acts" of grounding.*[5] The general idea is that the activity of one or a few words activates sections of the pre-frontal cortex to determine selective access and heterarchical activation in one or the other of the automatic grammar component networks. Intermediate reactions of the latter activate an initial sub-pattern of articulation. In quick processes of feed-forward and feedback interaction between grammar structure organizations and pre-conditions of articulation, the definite articulation activations can be sent to the distribution of articulation muscles. Obviously the processing stages flow from the pre-frontal top via intermediate morpho-grammatical grammar networks producing, in quick interactions, structure-appropriate activation patterns leading ultimately to bottom patterns of correct articulation specifications for the muscles, usually also involving smooth construction in the cerebellum. This example is a first sketch of pre-frontal cortices organization to be discussed in Chapter 4. The next chapter will moreover illustrate various aspects of prefrontally influenced organization involved in various types of concrete memorization and semantics.

Here is a final remark about the relevance of these processes in organizing "linguistic semantics." In addition to organizing common sense meaning there are still other interfaces that specify, internal to the semantic base, the use of thoughts/concepts to produce further thoughts/concepts for connections called "inference". They contribute to various types of "theoretical reasoning", also to making plans and forming intentions to act, thus forming "practical reasoning". There is moreover "social reasoning" that in some sense involves both theoretical moral reasoning and practical reasoning. Fuster also mentions that

[5] These linguistic operations will be explained in Chapter 6.

a comprehensive semantics must include various explanations of a richer and practically even more demanding set of boundary conditions that provide appropriate background knowledge for the various reasoning applications.

2.8 The ontogenetic formation of cognit memory

In our radically dynamic perspective we are not satisfied by describing changes of language competence, once they are already instantiated by neural connectivity. Instead we also want to understand first how connectivity-represented competence comes about in a basic but still schematic form in prenatal processes of the embryo and second how these schematic pre-structures are then adapted to the regularity acquirements of the language to be learned in the acquisition processes of the first 10 or 20 years of human life.

Let us see what can be learned from books about neurocognition. In the human neocortex, by the end of the second trimester of gestation, neuron generation seems to have been completed. But by the time the infant is born, some neurons are still developing, and axons are growing, which branch out and develop collaterals connecting with neighbour cells. Afterward, in the perinatal life of the human, as in all mammalian species, neurons, axons, dendrites and synapses undergo periods of exuberant growth and overproduction followed by attrition that is by reduction of their size and number. There is the plausible hypothesis of Changeux and Danchin (1976) that originally there is an overstock of synapses. After birth epigenetic factors related to neural usage "select" the synapses that will interconnect the neuron connections for the definite networks, while the rest of the overstock withers away.

In addition to these processes of developing cellular and connective architecture there are processes that develop connections to other structures, for instance to sub-cortical areas of the thalamus. Another important factor that contributes to long-distance connectivity is myelination of cortical fibres. In 1920 Flechsig observed that, in human development, cortical areas myelinate in a certain chronological order during the first years of life. In a later publication he concluded that the functions of the various cortical areas develop following the sequence of their myelination. This observation will be very important in our discussion of language acquisition in childhood. In any event maturation appears to progress from primary sensory and motor areas to areas of association. In pre-frontal cortex, maturation seems to continue until puberty.

In summary, the general genetic plan, the structural phenotype of the neocortex, is subject to a wide variety of internal and external influences. Through sensory–motor interactions with the external environment, the afferent, efferent and association fibres of the neocortex will develop and form the networks that properly serve cognitive functions. Fuster's summary is particularly instructive:

> There is a clearly genetic plan for the development of the entire observable structure of the neo-cortex. The plan covers all the macro- and microscopic features of that structure, including neurons and their connective appendices—dendrites, synapses, and axons. However, at every step of development the expression of that genetic plan, the structural phenotype of the neo-cortex, is subject to a wide variety of internal and external influences. Among the essential factors is the interaction of the organism with its environment. Through sensory and motor interactions with that environment, the afferent, efferent, and association fibres of the neo-cortex will develop and form the networks that are to serve cognitive functions. The development of these networks involves most likely a process of selection and neural elements among those that in earlier stages have been overproduced (selective stabilization). A degree of competition for inputs among cells and terminals is probably part of that selective process. Thus the elements that succeed in the competition would thrive and survive the normal attrition; others would be eliminated. It is a kind of Darwinian process. All the events of the neocortical ontogeny have their timetable (see Fuster 2003, p. 35).

3

Neural perspectives of semantics: examples of seeing, acting, memorizing, meaningful understanding, feeling and thought

3.1 Stages of complexity development in the perception–action system

The first chapter concentrated on principles concerning the global schemata of brain architecture and dynamic neural units. The second exposed the basic characteristics of the perception–action system in the mammalian cortex. I introduced a radical extension of Fuster's classical schema, complementing it by adding the organization components of language form perception together with structure-determined articulation action. This component also contains processing mechanisms that serve complex forms of higher order organization of formal meaning relations, intelligence, and creativity of art and so on. One usually assumes that these processes are located in Broca's and Wernicke's areas of the left hemisphere cortex. On the other hand concrete semantics and pragmatics may be distributed in almost any cortical area, locally or distributed, depending on the concrete meaning that the area or the area connection organizes.

In contrast to the previous chapters' schematic accounts of basically automatic perception–action organization, the present chapter will discuss a number of experimental brain studies concerned with special mental functions that marked breakthroughs in our understanding of processes in the cognitive cortex.

I think that an explanation of the functional systems and neurobiological architectures is more transparent when we do not directly consider the complete complexity of the adult's brain. Instead, we should study how the organization of competence develops in stages. Phenomenological observations of early phenomena and developmental stages of the brain's maturation during childhood and adolescence will help our understanding of functions and cortical processes. Let me indicate a few characteristics.

The developmental process begins with primitive experiences of the neonate, for instance hearing the rhythm of mother's spoken sound, seeing the configuration of mother's face and feeling the body contact when enfolded in mother's arms. The process continues to develop over many intermediate stages and leads finally to the competence of thought, reasoning and complex understanding of the social world of self and other selves, the natural world framework of objective things and events and its integration in a comprehensive framework of truth. Moreover, it may lead to personal developments and evaluations of even more comprehensive frameworks of moral, of world explanation and of fundamental reason.

The discussion of the earlier developmental stages is usually established in studies of developmental child psychology. Fuster (2003, p. 216) comments on the developmental stage schemata of Piaget's psychology as follows. It is of particular interest to follow these accounts, since they subsequently map the psychological content onto the neurocognitive framework. Fuster believes that already between birth and 2 years the child begins to form schematic networks of sensory–motor integration, that is, the stage of *motor reactions to objects* present in the child's immediate environment. Then *symbolic expressions* of acquired schemata begin to appear in the form of brief *stereotypical pantomimes*. The second stage, from 2 to 7 years, is the *representational stage*, in which the child extends the use of symbolism to the domain of the own verbal articulation, gradually understanding how words and word combinations relate abstractly symbolic features to the external world, supported by communicative understanding reactions of adults or grown up children. Manipulation of objects becomes progressively more regulated by feedback and by trial and error manipulation. The feedback includes progressively more language; in particular of the narrative and discourse integrative type supported by the logic of communicative play. The third stage, from 7 to 11, is that of *concrete operations* (skilful games and sports) in goal-directed and rule-controlled play and games, as well as by practical competence learning. This stage introduces quite a spectrum of reflexive reasoning. The fourth stage, from 11 to 15, is the stage of *formal operations*. Now the child begins to utilize hypothetical reasoning and to test alternatives. Both *inductive and deductive logic* flourish. Most importantly, the child becomes capable of temporally integrating information, and of constructing temporal gestalts of logical thought and action toward *distant* goals.

Evolutionary studies of complexity development are instead based on phylogenetic studies, in particular on measurements of non-human primates. In this range of studies the discovery of the *mirror systems* introduced surprising revelations. They concern meaning understanding of macaques, and thus explained the basically interdependent functions of perception and action systems. Section 3.3 will present some basic results of the mirror system studies of Rizzolatti and his group (2004).

3.2 Development of perception–action cycles

Let us now switch from the psychological story to the neurocognitive account. As already explained in Chapter 2, section 3, Fuster translates also the ideas of the former into his cortical perception–action cycle. The result is a hierarchy of neural structures that are dedicated to the integration of perceptual cognits and attention cognits that control executive cognitive actions. The developmental stages of the *child's intellect* readily suggest a recruitment of increasingly higher levels of *integration of cognitive networks* and *heterarchical connections* that allow selective access to sub-networks, simultaneously supporting inhibition of incompatible alternatives. We may think that, due to maturation, the neural substrate growth occurring at the appropriate age takes over the *integrative functions* of other substrates that supported cognition in previous stages. The new organization of higher-level efficiency may not necessarily involve the suppression of lower levels of competence. Instead, they become subordinated to the neural structures of the higher levels in the pursuit of higher goals. Whereas some constituents of those lower levels may be inhibited in this process, others may be used to contribute the integration of the more automatic actions to the higher gestalt of behaviour, language or logical thinking. Thus goal-directed action of progressively higher complexity is integrated at progressively higher levels of the perception-action cycle or by multi-modal connections to other areas at the same level. This is for instance the case in early word usage organised in Broca area whose location is not really high in the level sequence of the perception–action cycle. Later connections to higher semantically organizing Wernicke's areas and their semantic organization connection to a higher part of Broca's area lead to higher *level integration* as just described. Fuster adds some organization details: The feed-forward integration of those actions is assisted by continuous feedback signals from the environment through posterior (sensory) areas. Recall that, due to innate growth conditions, the *connections are usually reciprocal.* The pre-frontal cortex and its complex internal organization play an important role. It is late in maturation and becomes more efficient during the developmental stage between 11 and 15 years. Among the different functions organized by this cortex for instance is the temporal integrative role in the construction of novel plans of behaviour based on appropriate component selection and execution control. But its organization is also crucial for the integration of complex structures of behaviour, reasoning and language.

Let me now refer again to my model schema of Figure 2.3. It is a hierarchy showing a system of narrowly defined language form organizations parallel to the hierarchy of the non-language perception–action cycle. What would be the result if we compared Fuster's developmental account with Jackendoff's *developmental framework* to be discussed in connection with Figure 3.1?

I think that Fuster's description of the development of the neural hierarchy levels is suggestive. We would find some similarity to Jackendoff's sequence containing (a) phonological development, organized on the primary sensory level; (b) single word, word combination and simple syntax, organized on the primary levels of Broca–Wernicke connection; (c) morphosyntax, involving additional Wernicke's areas (near to inferior parietal areas); and finally (d) phrasal semantics and semantic-related grammatical function structure selectively guided and controlled by pre-frontal organization. In general Fuster's descriptions of developmental stages obviously apply.

But a fundamental problem is left. Linguistic grammar analysis is able to describe, for each word and for each sentence, the detailed structure constitution of the word's or the sentence's grammatical form. In contrast, there are limitations of neurocognitive science, It is not yet sufficiently developed to represent a precise network for cognit structure nor the precise usage condition that would activate the appropriate network activity of the given situation and context. The present methods of neurocognitive analysis do not yet provide detailed network description, nor does brain imaging map microscopic fine structure. Microelectrode measurements give access to single locations but not to areas or distant interdependencies. These facts do not exclude that before an availability of more precise technical methods analytic combinations of experiences and of specific measurements may provide empirically supported model constructions for system operations in the brain. The last chapter of the

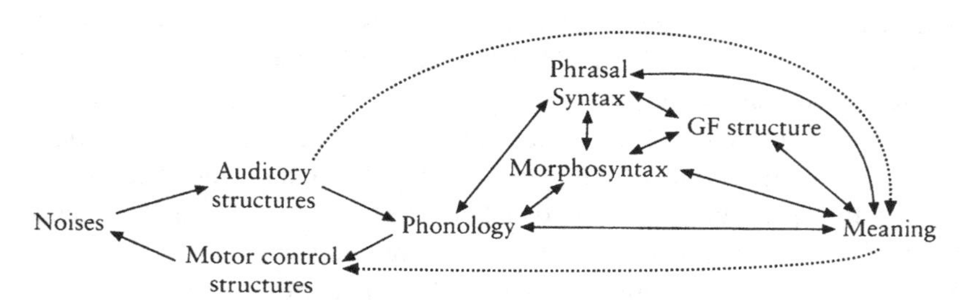

Figure 3.1. Grammatical components in a developmental architecture.

book will present and discuss an approach in which formal symbolic grammatical constructions are translated into models of operative networks in which local neural clusters and distributed interactions play a central role of organization. The method to be presented intends to satisfy Fuster's model criterion for binding processes. According to this criterion "information is encoded by the architecture of the networks that are translation results that normally are distributed and connected pattern in brain space; the firing regularities derive from the activation of that architecture, that is, from the characteristics of the intrinsic circuitry". (Fuster 2003, p. 226)

3.3 Mirror systems and the understanding of a perception–action concept in primates

So far, we considered ontogenetic developments in children. Is it possible to analyse brain structure organization in evolutionary earlier brains? The research of Rizzolati and his research group (2004) suggests positive answers based on very conclusive experiments with macaques. They discovered the so-called mirror neuron system. Let us now study the problem and its solution.

Imagine two experiences: (a) You are seeing someone grasping an object, (b) you do the same act; you also grasp the same object when given to you. Obviously, your experiences are very different. The first is an act of *perception*. You are *seeing* a person's gestalt and certain movements of the limbs, but you do not feel the muscles of your arm or hand. In the second case *you feel yourself executing the act* and thus you feel your arm and hand moving. But in grasping you do not see an acting gestalt and the movements of the limbs. You do the grasping! It is clear that the two processes are completely different. In earlier times philosophers and scientists reflected about these cases. Some philosophers and psychologists have argued that the sameness of action, the action of grasping, is learned by language. It was claimed that understanding that seeing someone else grasping and doing the grasping yourself are the same actions is learned from the adults' using the same word grasp, whether, doing or seeing. If this explanation were true one would have to conclude that animals and even the non-human primates should not be able to know the sameness, the abstract concept, of grasping because they cannot learn a language. In this behavioural tradition action and perception must be considered as strictly different. Moreover, neurologists have shown that perceiving the act and executing the same act and are organized in strictly different brain areas, the former in the posterior part of the brain, the latter in the frontal part.

Rizzolatti and his group have shown by measurements in macaque brains that, among neurons in the frontal brain, some activate the muscle movements of grasping. There are a few that are also activated when the macaque sees someone grasping (1995). More precisely, if a neuroscientist grasps a berry and presents it to the macaque and the macaque immediately grasps the presented berry, the specific neuron is continually active during the perception phases and in the immediately following phase of the own grasping. Rizzolatti could also show that, during the first phase, the same neurons are active. The *mirror neurons*, as Rizzolatti called them, signalize a mutual relation between the frontal brain activity that activates action and the posterior brain activity of perception. He could also show that a neural connection between posterior and frontal activity provides the base for the simultaneous indication of grasping in one or the other form.

This is obviously similar to the synchronous activity based on mutual interaction that Fuster integrated in his modification of Lichtheim's model (see Chapter 1, section 7). But note what Rizzolatti discovered: There are integrations of many pairs of activity in perception areas and activity in action areas that correspond to each other because there is a *functionally efficient fibre connection*. Consider the following statements referring to the macaque experiment:

(p1) Jack sees Jim grasping a berry. (m2) Jack grasps the berry.

Let the statements (p1) and (m2) express neural activities. (P1) is the process that exists in Jack's brain as part of a network that organizes the types of perception while network (M2) in Jack's brain organizes the act of grasping. Let Jack be the name of the macaque. In a more specific explanation (M2) is primarily a mirror neuron of the same type as the one discovered by Rizzolatti and (P1) is primarily the perception area identifying the grasping perception.

In recent years the studies of the Rizzolatti group made important progress. Among other aspects it could be demonstrated that mirror neurons also exist in human brains.

In the linguistic perspective it may be interesting that Rizzolatti and coworkers (2000) introduced a sort of "vocabulary" for motor acts related to prehension. The intention was that the word differentiation should characterize different experiments. Thus "words" should denote populations of neurons related to different motor acts. We might also say that the words should guide the search for different cognit networks. The selection contains for instance the terms *"grasp," "hold"* and *"tear"* thus distinguishing kinds of acts. Other terms, such as *"precision grip", "finger prehension"* and *"whole hand prehension"* denote cognit networks that differentiate more specific types of grasping execution.

Another complex of Rizzolatti's studies addressed the question whether specific evolution of mirror neurons in our species may have contributed to our early ancestors' *development of language* (Ferrari et al. 2003). Recent studies have shown that most neurons in the lower part of the mirror neuron area, which is near to Broca's areas in human brain, are indeed not specialized in organizing grasping acts. Instead they organize mouth movements, some responding to visual perception of ingestive actions, others to communicative mouth movements.

In another perspective Rizzolatti and Craighero (2004) argued that one of the important roles of a specially developed mirror neuron system is identifying not the action per se but the *goal* of actions done or partially indicated by others. Thus it may be said that mirror systems transform visual information of an action into *goal identification,* serving as a determining part of executive knowledge.

Close cooperation with the Rizzolatti led Arbib to systematic reflections about language evolution (Arbib 2002, 2003). Reflecting about the assumption that the human brain's architecture is "language ready" Arbib considered the functions that are near to the human competence or only accessible to human competence. Here is his tentative list:

> Symbolization: *The ability to associate an arbitrary perceptual symbol with a class of events, objects, or actions etc. At first these symbols may not have been words in the modern sense, nor need they have to be vocalized.*
>
> Intentionality of articulations and gestures: *Communication is intended by the speaker to have a particular effect on the recipient.*
>
> Parity (which Arbib derives from the property of the mirror system): *What counts for the speaker (or producer) must count for the listener (or receiver).*
>
> Hierarchical structuring: *Production and recognition of components with sub-parts. This relates to basic mechanisms of action-oriented perception with no necessary link to the ability to communicate about these components and their relationships.*
>
> Temporal ordering: *Temporal activity coding these hierarchical structures.*
>
> Beyond the here-and-now: *The ability to recall past events or imagine future ones.*
>
> Paedomorphy and sociality: *The prolonged immaturity of the infant and the prolonged care-giving of adults combine to create conditions for complex and social learning.*

Arbib emphasizes that the first three functions support primate communication systems without necessarily yielding language while the last four properties are particularly relevant also for the development of language complexity. For instance structure-accessible hierarchical structuring with temporal ordering is particularly important. Automatic visuo-motor usage is certainly already accessible in macaques. But it seems that they are unable to distinguish the whole and the combination groups in the hierarchy system. That is, they are unable to "syntactically parse". The two final criteria are even more specific for human development. They contribute in particular to the typical development of language competence.

Arbib's further reflections relate also to the grammatical aspect organizations we discussed in the previous chapters. He emphasizes that the purely *automatic organization of simple hierarchy combinations* becomes modified as soon as pre-frontal attention organization acquires *heterarchical access* to subunits together with reorganization of earlier and simpler structure systems. It is probably true that the first *Homo sapiens* used a form of vocal communication which was but a pale approximation of the richness of current languages, and that these languages evolved culturally as an increasingly cumulative set of "inventions" based mainly on the last three criteria. This does not contradict the existence of certain innate competences. They imply specific restrictions. Inventions must be communicatively efficient and must not conflict with correspondences of speakers' and hearers' brains, a criterion that Arbib calls parity – the property "*What counts for the speaker must count for the listener*." This is obviously an extension of the original mirror system phenomenon according to which the features that count for the execution of grasping must count for the visual perception of another's grasp.

This is Arbib's summary: "Biological evolution equipped early humans with 'language-ready brains' which proved rich enough to support the *cultural evolution* of human languages in all their communalities and diversities." More specifically, Arbib argues, "that what *Homo sapiens* possessed was not protolanguage... what they possessed was the ability to name events with novel sequences of (manual or vocal) gestures, but that this capability does not imply the ability to separately name the objects and actions that comprised those events. The latter ability was a momentous discovery made by humans perhaps 100,000 to 50,000 years ago, rather than a biological heritage from earlier hominids" (Arbib 2002).

I think that Arbib's specifications are indeed very interesting. They share, however, a typical shortcoming with other proposals. The analyses are strictly related to aspects of perception and action, and thus to the cortex. They do not consider the role of the complete nervous system involving emotions and feelings and the combinations with the body's somatic organizations combining word sounds with dance and music that were certainly important in early human rituals. We must assume that such rituals – mainly executed in contexts of stone configurations and painting – were expected to influence conditions of daily action and also rituals that were connected with death, whether accidental or ritually intended. I do not have any doubt that the combinations of dance, music and words, were well known before ancient Greece. Based on ideas of transcendence that had important status these contexts strongly contributed to language development early in prehistoric times.

3.4 Measurements of stages in children's language acquisition

Let us now consider another domain of empirical studies, those determined by brain imaging.

Here, again we are most interested by those aspects that relate to the models of the previous chapters. Measurements by Friederici have indeed demonstrated that auditory sentence processing is hierarchically structured in time and in the brain's topography. As already argued several times a number of different cognit networks must be involved. They become activated in organization processes. Their interaction generates a complete unit of understanding, resulting from a binding process, probably resulting from interactive integration. Theoretical reflections of linguists have suggested that sentence processing proceeds in steps: First the hierarchy of words and word combinations and then their interpretation in terms of meaning. Another model based on psychological data claimed that syntactic and semantic processes interact from an early stage during auditory language comprehension. Friederici's measurements (2002) prove that both are correct at their times. The timing of the brain processes activated when hearing a sentence involves only language form, that is phonology and word form and category identification, during the first 200 msec, hence in the first common sense moment. During 300 to 500 msecs analyses of semantic categories and combinations become involved parallel to checking syntactic structure. At about 600 msec syntactic structure and semantic meaning are integrated.

These observations of hierarchical organization have been confirmed by behavioural studies in infants (Dehaene-Lambertz et al. 2008, p. 405). But measurements of infants did also show that their longer processing time or more sustained activity might be the result of different cognitive operations that integrate over increasingly larger areas representing more abstract – or combined – speech cognits. The authors argue that such a nested organization of processing units with a progressively longer temporal window of integration would provide infants with an adequate tool to segment the speech stream into its prosodic components.

The research group's functional magnetic resonance imaging (fMRI) results also indicate that the brain regions that are involved in receptive speech processing in infants are not limited to unimodal auditory regions. They extend to remote regions, including areas such as the frontal regions. This can be demonstrated from the first weeks of life onwards. The human brain displays phonetic categorization capacities, rhythmic and prosodic sensitivity, which make it particularly adapted to processing speech. These capacities mostly rely on brain circuits close to those observed in adults, i.e., the left perisylvian areas. The similarity between functionally immature infants and competent mature adults implies a strong genetic bias for speech processing in those areas. The functional properties of the superior temporal areas and their connectivity with remote regions in the humans might be crucial to ensure language learning.

Further details come from recent fMRI studies of 6-year-old children in Friederici's group (Brauer et al. 2008). They demonstrated that children's responses showed overall longer latencies when compared to adults. It is surprising that they found a temporal primacy of right over left hemispheric activation. Longer temporal delays, when compared with adults, are in line with the group's current understanding of maturational changes in language-related brain areas and the structural connections between them. The data also support the view that developmental changes evolve from higher processing costs in the developing brains of 6-year-old children to faster and more automatic language processing in the mature brain.

Particularly interesting are the observations that the brain's maturation has influence on complexity competence. The authors consider the following scenario regarding functional and structural contributions to the development of language comprehension. The overall differences of measurement exist mainly due to ongoing maturational changes in children, whereas specific age differences between particular brain areas might be mainly based on differences in functional processing in which structural properties contribute less. These are clear cases of different processing strategies in children and adults. I think that Pulvermüller (2002) and others' observations are correct that babbling is essential for basic organizations for building up language-specific neuronal representations. In children prosodic foundations rely more than in adults on right hemispheric organization, probably they are more related to communication and situation evaluation that is typical for the right hemisphere. In general these observations might suggest that, as long as children's brains do not possess mature connectivity for advanced structural organizations, they need to compensate for that disadvantage by momentary strategy and or effort. Further brain development (through maturation and experience) will allow more effective information transmission and processing (Givón 1995).

3.5 Visual and auditory parts and wholes in the brain's space and time

We now leave the discussion of measurements and observations that might contribute to a better understanding of language evolution and development in our species and in individuals. We now turn to observations and analyses of adults perceptions. Their organization in the brain involves a number of processes that are surprising because the basic phenomena are not directly accessible to our own conscious experience of the details of vision and audition. When hearing a fluent sentence utterance in our own language, our psychic act identifies the sound pattern and the grammatical groupings of parts. The organization of acoustic details, or details of grammatical correctness, usually remain unconscious except when they indicate a special dialect or there are grammatical errors. Though non-conscious, minimal differences and errors are signalled. If, however, the sentence refers to a concrete visual object or a situation, the analysis of visual perception can demonstrate that essential components of usual organization remain strictly unconscious. Measurements show that complex brain processes of properly and efficiently executing eye-movements are involved.

In the following discussions I will first concentrate on visual and tactile perceptions often connected with action execution of hands, arms, legs etc. If they indicate objects and situations these are often understood as entities of *visual and or gesture-indicated gestalt or configuration. Our conscious experience of near and accessible objects is that of a spatial configuration.* How does it come about?

Let us first consider a situation in which vision is part of a communicative situation in which *visual- or haptic gestalt- or configuration*-features of objects and situation moments are expressed by words, phrases or sentences. The objects may be indicated by pointing or by form specification initiating brain processes to identify *spatial gestalts of objects and events*. What is seen appears to be a momentary configuration of static things. This seems to contrast with the experience of the spoken words and sentences; they are not static events but *temporal gestalts* of sound events. When reading a text we are also confronted with static gestalts of letter configurations. But in the case of sentences the brain immediately memorizes not a letter configuration but a corresponding auditory sound pattern. This concerns the sentences and the discourse meaning. It does not exclude the strange fact that experienced readers of books can recall that this or that important statement was printed on a left side page in the upper part. This proves that the configuration of certain text elements is still located in the experimental space.

But all of what we said relies on common sense understanding. Its content is not really justified by experiments and studies of mental behaviour or details of the brain's organization. This and the following sub-sections will explain some interesting details.

3.5.1 Saccadic eye movements and possible counterparts in hearing and articulation.

The surprising phenomena of visual perception are connected with spontaneously and automatically organized processes of *saccadic eye movements.* I will first concentrate on typical cases of visions of objects and situations. Their role in reading as well as some sequential counterparts in concentrated speech will be left to the last part of the section. Presently the primary interest is to correct a number of errors in our common sense understanding of vision and audition. I will show that the difference of common sense assumptions and real organization of vision derives from the fact that certain core processes, the organizations of so-called *saccadic eye movements* which selectively execute visual situation scanning, are quicker than momentary common sense experience. In the visual process there is a continuous cooperation of the saccadic movements with identifications of features given at a focused point and their integration with features at other focus points. Together the different operations that generate a single eye movement with focusing moment take about 200 milliseconds and are thus slightly shorter than our common sense momentary experience.

Discussing a phenomenon of a simple example may be particularly useful. The phenomenon was often used for illustrating the binding problem[1] characterized as a fundamental problem that would be difficult to overcome by standard types of computational connectionist spreading activation models. Jackendoff's version is as follows: "We have found that the shape and the colour of an object are encoded in different regions of the brain and they can be differentially impaired by brain damage. How is it, then, that we sense a particular shape and colour as attributes of the same object? The problem becomes more pointed in a two-object situation: If the shape region detects a square and a circle, and the colour region detects red and blue, how does the brain encode that one is seeing, say, a red square and a blue circle rather than the other way around?" (Jackendoff 2002, p. 59).

It is obvious that here the underlying model of visual feature perception assumes that the brain *registers simultaneously* all visual features in the visual

[1] See above Chapter 2, section 3.

field. Moreover it is assumed that each possible combination of active features is registered by a momentary activity at its specific location. Since blue, red, circle, and square are perceptually activated features, there are four activated combination locations. If all combinations were possible in the brain the system would signal that the circle is blue *and* red and that the red object simultaneously a circle and a square. The error in this argument is that generally all activated simple features are simultaneously combined in the posterior cortex. The role of the scanning movement of the visual fovea and the registration of their sequence positions are completely left out of consideration. But in fact the pre-frontal cortex organizes the focusing sequences and asks, at any focus moment at the posterior cortex, which are the activated feature signals *at the momentary focus point*. The answer, for the focus point, will be a distinctive feature activity, in our case one for colour and one for shape. The argumentative error is that the role of the focus scanning of the pre-frontal cortex – cooperating with the superior colliculi – is completely neglected.

Returning to our example we understand that the situation is extremely simple. Here is a simple repetition of what happens: If the square and circle in the extended visual field are small and the eyes are initially focused between the two, the next move might lead left to focus the red square whose features are immediately registered by the visual system. While being at the focus point the selective brain determines another interesting focus point in the extra fovea retinal periphery to the right. The move is executed, the next focus point is reached, and the focused data, namely the blue colour and circle form, are registered. Thus the visual system summarizes: Left of a middle focus point there is a *red square* and to the right there is a *blue circle*. Registering a blue square and a red circle at the same focus point is impossible since no colour-form combination of this type can be registered at any momentary focus point.

The basic error of Jackendoff's exemplification is the assumption that all features red, blue, square, and circle can be simultaneously focused. The situation is different for a camera: The release of the camera shutter opens a parallel optic data flow whose detail distribution is indeed simultaneously registered. But the eye's separation of fovea and retina periphery and its sequence of fixation points don't allow this. Instead the eyes *can move and scan* the visual field by a sequence of focus points. Thus the analytic errors consist, first, in neglecting the binding of the "feature pair" fovea-focus *position* and *fixation time interval* with the concrete perceptual features like colour and form elements and, second, that distinct colour-features and distinct form-features cannot simultaneously be true at a given focus-position during a fixation moment.

3.5.2 Perception and action in organizing vision and speech

Saccadic eye movements are surprising because everybody's eyes show these movements but nobody experiences consciously the momentary changes of the saccadic movement fixations and the fixation sequence while spontaneously perceiving an object or the objects of a situation. The reason is that the *single saccadic eye movement is not intentional* nor is the eye movement executed at the short interval of the fixation moment. The general intention of the latter is to get a momentarily sufficient constituent feature possibly relevant for the situation. Definitely the elementary features fixed by the momentary eye movements are quick; each registration is executed in about a quarter or a third of a second.

Let us consider the process. Suppose you notice an object that somehow arouses your interest. A brief global impression activates your mind's attention focus and causes your intention to identify clearer what is behind the impression. Without further intention your eyes will immediately move so that the object or an object feature falls on the fovea, the position of the retina where vision is most acute. But this is merely the *core of the impression*. A clearer identification requires *information about the neighbouring positions* whose features might help you to understand the complete phenomenon. Saccadic eye movements will focus one position after the other. The sequence of focus points will be called an f-sequence.

So far the description is merely a story. But several mechanisms for measuring and representing eye movements have been developed and applied in vision analyses. Let me concentrate on a typical measurement as presented in Figure 3.2. It presents the f-sequence of fovea fixation points. The movement from fixation point to a potentially relevant next fixation point represents the saccadic eye movement.

3.5.3 Brief explanation of functions that the brain organizes in connection with saccadic eye movements

Figure 3.2 presents merely the fixation points and movement direction of the f-sequence. Let me explain what happens at the points, by referring to the f-sequence in Figure 3.2. Initially the central body part of the fox is focused. Next there is a short indecisive move to a nearby position. Then a clear sequence starts: fox's neck, mouth, eyes, ears, four legs, three tail positions, two different positions at the hind legs, returning back, near to the original body centre position. In this story the fixation positions were named. If there is a human observer it might be possible that he imagines the names simultaneously with each

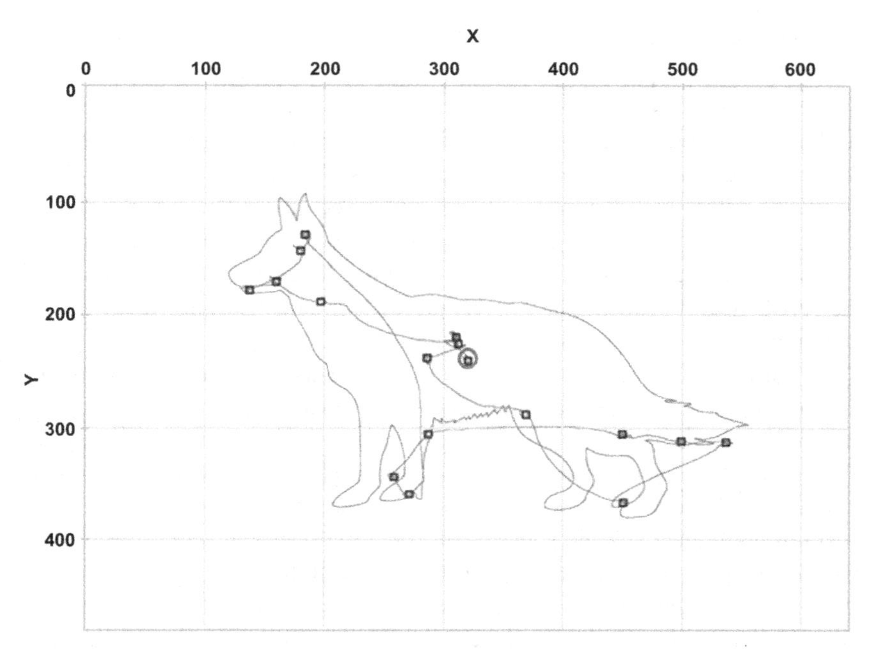

Figure 3.2. Saccadic eye movements – Scanning a design of a fox.

fixation position. Obviously, in an animal, perhaps a cat looking at some distance at the fox, the cat could not generate words in mind. But it may also have different visual category cognits that activate at the moment when they fit the fixated feature.

Note that we implicitly introduced a new function in addition to those previously mentioned. Let us call it *saccadic integration*. Neighbouring fixation positions are checked for applicability of common categorizing. There are some neighbouring sub-sequences in our example such as those for mouth, eyes and legs. We do not see this leg and that leg as completely different objects; rather they are perceived as the fixation bundle of legs. The same holds for eyes, ears etc. It is probable that these combination categories are immediately registered when the scanning shifts are still in process.

There are further integrations. One concerns the end of the spontaneous f-sequence. Usually, at the end of the f-sequence the function of integrating category registration requires a "summary" category characterizing the object or the situation type, perhaps together with its practical and or emotional evaluation of behavioural relevance. For instance, if the brain of a cat organized our f-sequence, the result might activate perceptual, emotion and attention categories like: "Fox! Not moving my direction! Still: Better to take care!" The brain is most obviously capable of "summarizing" the whole situation suggested by the f-sequence combinations of features. It must even activate a learned dynamic state expressing the "feeling" of the whole as an object figure or the configuration of an understandable situation. Since an f-sequence is a discrete sequence of fixations, the integration is obviously not merely a summation of pieces in a continuum (Baars 2007a sect. 2-1). The competence of generating the integration is part of "common" sense. The integration features can be prepared for long-term memory fixation. Normally we do not register the details of an f-sequence as discrete pieces of momentary knowledge but as an integrated object or a coherent situation.

All of this is surprising when we consider the speed of the brain's operations during scan path generation. Let me briefly mention the temporal characteristic of the f-sequence. Each eye movement needs 200 or 300 milliseconds. The complete f-sequence of our "fox" example takes 4 to 5 seconds. But it is clear that each movement has two components: the attention search for the next position preference and the subsequent execution of the directed movement. Each of the former takes about 50 milliseconds and the proper eye-movement 200 to 250 milliseconds. All of this is very quick and temporally shorter than a common sense moment. It is however important to register that the brain must execute already two operations, the search for the next attention shift and then, when selected, the execution of the next movement.

Given the characteristic status of saccadic eye movements it may be interesting to contrast the single object scan with a situation scan of the configuration of objects.

In Figure 3.3 the measuring procedure represents also the length of the time interval of each fixation by the size of the circle around the fixation position. As in the case of the scanning of the fox design the eye movements are also unconscious. If asked about their experience test persons will just mention the objects and their arrangements, but not their eye movements. If several objects of the same type are present there is no danger of confounding the tokens. What is the reason? The answer is easy. Not only the feature characteristics are identified at the fixation. If this were the case the two plates could not be distinguished. Direction and distance are the additional information that is registered. Thus, distinguishing tokens is easy since the token appearance is always connected with the individual or relational location features.

Before concluding the sub-section we should briefly add a feature that may become relevant in more complex procedures that do not directly reach a successful termination. Even for the simplest case of Jackendoff's red square and blue circle Treisman (1988) mentioned, "that under time pressure, subjects can mismatch the features of multiple perceived objects". But registering or mem-

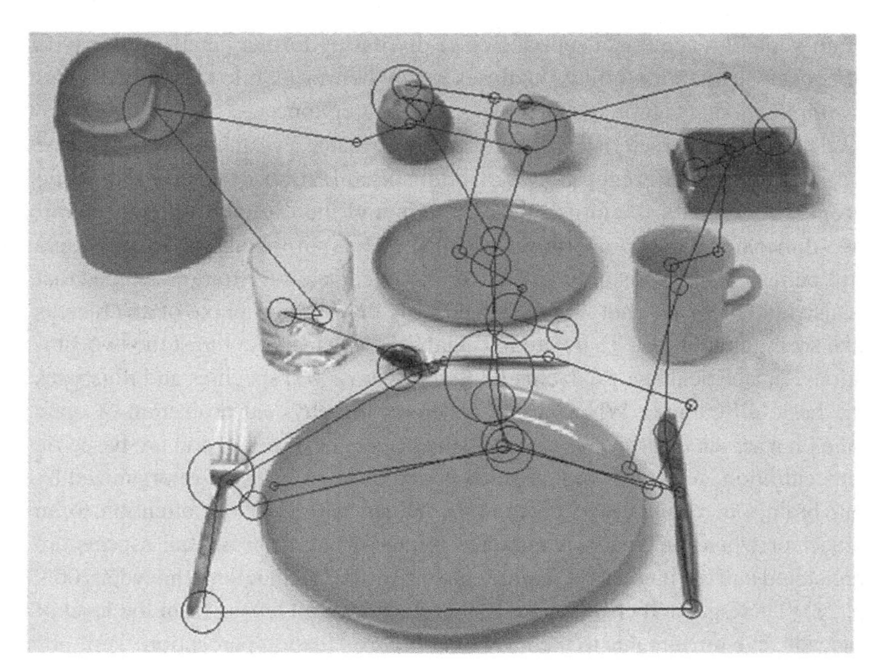

Figure 3.3. Saccadic eye movements – Objects on the table.

orizing in the example case of a blue square and a red circle should not lead to error at time pressure. More complex are the situations in which an f-sequence reaches a moment of confusion, for instance because the complete set of the fixation point information is not sufficient for a proper identification of an object's or a situation's configuration. This would disturb the selection process that is necessary for construction of the f-sequence. For instance the visual field may have produced plausible pieces of figure components but together they do not easily fit together or can be composed into a plausible relation. This is often the case in optical illusions and particularly in cases of distinguishing objects in the foreground from those in background.[2] The visual process of fixations starts with intending to construct a particular object interpretation that is subsequently contradicted by other features. The optical illusion problems are different from those that were previously described. They activate more attention concentration in a procedure that has characteristics which are similar to those described by Fuster (2003) in his chapters about the

[2] See the particular examples in Schnelle-Schneyder (2003).

brain's intelligence organization. Before discussing further details of this type of conscious attention-guided analyses it may be worthwhile to briefly discuss its implications for more general aspects of perception.

The visual characteristics illustrated by saccadic eye-movements indicate a fundamental difference between brain organization of vision and optic processing of data fixations in the camera and the digital computations of two-dimensional data properties in computers. It is surprising that many people still believe that our vision is in fact similar to camera registration, an idea that is also held by many photographers, believing that taking a photo of an object is like seeing that object. There are few analysts that clearly contrast the two situations. In a critical analysis Schnelle-Schneyder (2003) specifies and illustrates the basic differences. Whereas one generates the physical projection of optic data on a screen or projecting surface, the principles of our vision are based on concentration, selection, and abstraction and these principles are organized by our brain's mechanisms of perception. "When we direct our attention to an object or when our vision is attracted by an object, the essential aspects are registered and the inessential features are neglected" (Schnelle-Schneyder 2003, p. 22). These general characteristics are relevant on all levels, from the level of saccadic eye movements to the level of complex situation perception.

3.5.4 Cortical and sub-cortical brain components and their cooperation in spontaneous vision processes

As promised above I will now turn to a discussion of the *cooperating brain components*. Let me start with a story about brain processing during generating f-sequences of vision. The "acting figures" of the story are brain areas and nuclei. Though there are several areas involved I will only mention the most important ones. Namely the *posterior eye field* (PEF), the *frontal eye field* (FEF), the *dorso-lateral pre-frontal cortex* (DLPFC) and finally the *superior colliculi* in the mid-brain. The local positions of the first three are named by abbreviations in Figure 3.4, whereas the small location of the colliculi is indicated by the line from the lowest right-hand box in the figure. The figure shows that many connections can be mutually active in both directions.

My interpretation of saccade-generating processes begins when the eyes focus for a moment on some central position of the environment, based on cooperating procedures in the frontal eye field (FEF). The registered features are signalled to the superior colliculi. Salient perception information in the posterior eye field (PEF) activates an eye movement signal at the FEF and the superior colliculi. A sequence of new salient information will generate a focused sequence.

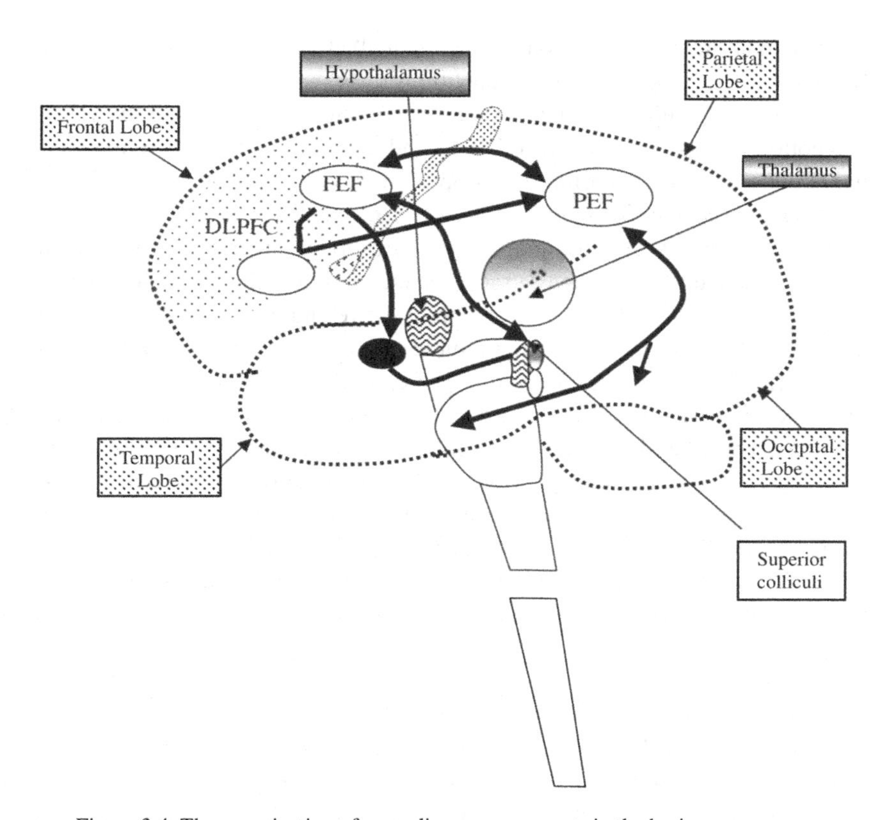

Figure 3.4. The organization of saccadic eye movements in the brain.

Note however this cooperation does not yet exclude a problem. It would consist in several repetitions of the same salient focus point. Avoiding this problem requires additional cooperations. Assume that during a focus sequence the PEF and FEF together perceive again a feature that was perceived at an earlier moment during the sequence of focus positions. It now becomes important that the FEF and DLPFC cooperate. The latter produces functionally efficient excitatory and inhibitory signals. When there is no contrary situation expressed by inhibitory signals from the DLPFC the eye movement to the salient position can be activated after the 50 milliseconds usually available for a momentary selective operation. The situation is different when the FEF and DLPFC have received contrary information about the salient feature from the superior colliluli. In fact it is one of the tasks of the superior colliculi to *register all earlier fixation positions* in the colliculi's *internal space map*. Signalling them back to the frontal cortex *causes inhibitory signals* from the DLPFC in the case that an earlier position is again signalled as salient. In this way the focus sequence does not generate repetitions.

In view of feature identification the system must have access to the visual perception processing. As Petrides and Pandya (2002) explain there is moreover cooperation with the inferior temporal region that stores the *knowledge about visual specifics of objects and situations*.

It is particularly easy to exemplify the categorical integration as discussed with reference to the fox picture saccades (see Figure 3.2). The first saccadic shifts started from the fixation on the body centre and lead to two fixations at the mouth. In the categorical integration act the two fixations are marked by a cognit that represents the integrative perceptual category mouth. The same principle of marking sub-sequences will be applied on subsequent saccadic fixations. Thus the identification categories eyes and ears and still others, the front legs, tail etc. become represented by different internal cognit activations. In this way most sub-sequences are integrated by perceptual category cognits. If the observer is a cat, its brain registers the perceptual category cognits just named at the appropriate fixation sub-sequence and finally the complete characteristic feature of the perceived objects or situations. The cat would react as suggested in the discussion of Figure 3.2.

If however the observer is not merely an animal but a *human adult* and a *speaker* of a language not only summary categories in the perception–action cycle are activated. There are also connections to the areas of language form processing, that is in *networks of LF-cognits*, in which appropriate words and word categories are activated and brought into a *linguistic form-meaning binding complex*. Applying this observation on the saccadic perception of a mouth, the sound experience of "mouth" will become cooperatively synchronized in a human adult but not in an animal.

Let us note a last point. The operations of the eye movement system generate primarily a *temporal* sequence of foveal fixations, whereas the result is memorized in a *spatial* visual representation, a gestalt. How does the brain compute the *perceptual* presence of a *spatial gestalt*? Could we start from considering the integrative components fox, mouth, eyes, ears, legs? I think that there is an additional connection of the areas mentioned so far with the cortical somato-sensory areas, namely those that represent an *internal feeling* of an arm-and-hand movement over the fox's fur. I therefore assume that a connection with these somato-sensory areas is very important in the brain's generation of a spatial gestalt. It is possible and even probable that the relation to these spatially relevant areas, the arm-hand sections of frontal parietal and of pre-motor areas, are integrated in a visual process described. That is, not only integrative categories are registered but also simultaneous touch movements determining a *spatial gestalt feeling*. But so far I could not find out whether this is true. The discussion of systematic aspects of category assignments to parts and wholes in vision and also in sub-sequences of sound patterns will be taken up again in more theoretical and systematic form in the last chapter of this book.

3.5.5 Eye-movements in acts of reading

As an annex to normal vision organization I shall discuss the visual act of reading printed texts. From a superficial point of view written texts are visual patterns and their scanning should be somehow similar to looking at pictures. And indeed, reading a text involves a series of eye fixations separated by saccadic eye movements. But it is clear that attention selection of fixation points is obviously simpler in a horizontal reading act than in visual scanning of objects and situations. The simplicity of eye-movement will be illustrated by the following schemata presenting the visual reading stages of the sentence "*Graphology mens personality diagnosis from handwriting.*" (Cp. D. Caplan 1999).

~~Grap~~hology **means** pers~~onality diagnosis from hand writing. This is~~
~~Graphology means~~ **personality** ~~diagnosis from hand writing. This is~~
~~Graphology means personality d~~iagnosis **from** hand ~~writing. This is~~
~~Graphology means personality diagnosis f~~rom **hand writing**. ~~This is~~

Each line presents a fixation position of each of the four saccades. The bold face words mark the left to right movement of the fixation centres and the sections presented by stroked letters are in the periphery of focal vision.

Three important empirical facts have emerged from analyses of these measures. First, eye movement patterns are exquisitely sensitive to linguistic properties of the words being fixated and how the words fit together to make grammatical sentences tell a sensible story. Thus the measurements of reading complete sentences in a text allow more fluent reading than measurements in which the test person is forced to read and understand the meanings of a sequence of isolated single words. Forced single word reading is obviously much slower than the context coherent words and sentences. Caplan explains the measurement details for fluent text reading, whereas Posner and Raichle (1997) concentrate on measurements of single word readings. The reading situations are different since the useful information obtained during any fixation in continuous text comes from at most a couple of words beyond the word currently being fixated. Function words (articles, prepositions, conjunctions etc.) are often skipped in eye movements whereas content words (nouns, verbs, adjectives, adverbs) are normally fixated. Longer content words may receive two or more fixations. They are rarely skipped. The temporal course of fixations is similar to f-sequences in two- or three-dimensional vision. Both are sequences, straight in print but curved in situations. The average of a reading fixation takes about 250 to 275 milliseconds. Thus the brain must very rapidly implement the different brain computations of different sub-tasks in determining sequences of reading fixations.

The previous remarks about text reading refer to word sequences that occur in well-known syntactic composition and in semantically meaningful context. We must expect that the pure identification of words and phrases in moments of fixation interact with other online processes that check the correctness, meaning and background appropriateness. The result is the integration of letters and meanings in text understanding. Friederici's (2002) measurements of syntax organization present the details of several stages. The last step may indicate the integrative completion step.

3.5.6 Verbal rehearsal

When we now turn to the study of verbal rehearsal we may wonder whether here also several features appear as combined into cluster representations. In connection with his studies of verbal rehearsal Baddeley (2002) proposed a model according to which rehearsal organization is operated upon in a working memory that would store data to be rehearsed. Baars adapted the proposal in a neural model: A working memory is a system of the following interactive components: (a) the *central executive* of the pre-frontal cortex containing the more general and abstract cognits, (b) the *working "storage"* or *buffer* providing access to the more concrete and specific cognits, and (c) what Baars calls the "*inner senses*". The competence of "*verbal rehearsal*" is one of the inner senses.[3]

Let us briefly consider Fuster's (2003, p. 130–131) critique of the classical view that is still represented in Baars (2007a, p. 32–33). Fuster (2002, p. 196) writes that 'reverberation through recurrent neuronal circuits is a likely mechanism of working memory and therefore of *temporal integration*. Consequently working memory appears to be a *mechanism of temporal integration* based on the *recurrent activation of cell assemblies* in cortical long-term memory networks.[4]

In our functional perspective verbal rehearsal consists of certain kinds of mental *access to linguistic cognits* by activating them in immediate memory. The classical example, which everybody can easily imagine introspectively, is

[3] See also the more recent summary of Baddeley (2002).
[4] I will return to a more detailed discussion in Chapter 8.

the act of trying to remember a telephone number or a shopping list. In the first case it is obvious that the brain contains a sub-system of names for the primitive numbers in their ordinal sequence from 0 to 9. The learner can identify the names in reading or hearing, followed by memorizing the names in the sequence in which they are presented for a short moment. The moment is sufficient for initiating repetitions. Repetitions stabilize and memorize the knowledge of the number sequence for a period. The sequence is repeated again several times, in view of better stabilization of knowledge.

In addition to sequence learning, which uses short time memorization, the learning process may also use additional ways of memorizing. If the number is 23457846 the learner may separate the eight number sequence into 2345 78 46, recalling the first four in their normal ordinal number sequence and memorizing the following sequence in number pairs. Some learners may even use additional features of memory support. They may memorize that the sequence could be based on 2345678 4 finally requiring in addition that the number 6 must be moved to the end.

In any case there are various learning techniques relying either directly on simpler parts in the focus sequence, or on some grouping of sub-sequences or on movement operations etc. It is clear that everybody has learned already as an infant acts of grouping babbling sounds, as well as later, hearing and recalling repeatedly songs and poems heard from the adults. In this case, as well as in first learning constructions of word combinations and sentences, even groupings of groups are learned. They lead to something like an understanding of a primitive form of structural hierarchy in grammar.

Often the intention of the *psychologists' perspective* was to select separated and usually unstructured forms of single and specific functions as components involved in the integrative process. Early studies concentrated on sequences of words or items in which each unit is unpredictable in the context of the others. These studies led to the famous result that the limit of a short-term memory buffer is about "seven plus minus two" separate items. In order to motivate the test persons one usually says that their task is similar to the task of recalling a telephone number. Obviously, such an indication is necessary in order to provide the test person with a background that motivates the test.

In a superficial interpretation the test may look similar to the process of saccadic eye movements. These also produce a sequence of fixation points each with a visual fixation feature that attracted the eye movement. But remember that during the constitution of the focus sequence, also combinations of positions were integrated in complex features. This was possible because the process made use of *background knowledge* about the gestalt elements of an animal. This led to a hierarchy from local to more combined features.

What is new? The rehearsal process accounts for a *background knowledge*: The fact that the set of numbers is an ordered set. Thus the first four numbers of the sequence repeats the order, starting with the number 2. Concentrating in the act of memorizing the learner discovers that memorizing is easier when discovering and *using an organizing rule*. As our example shows, the rule can be discovered during the rehearsal process. Consider for instance the sequence 23455432. One easily discovers that the second half is the inverted sequence of the first half. Or take 23454523 where the sequence of the first two pairs is inverted. In all cases, one is able to discover a more regular order specification, supported by background knowledge. Thus memorizing is easier when based on *knowledge of grouping and structures*.

In considering such aspects Baddeley was led to introduce an additional organization principle, available in a unit called *episodic buffer*. One major feature of the episodic buffer concept is its emphasis on the important issue of how *information is chunked*, how it is related to the more general concept of binding and how it can best be studied. (Baddeley 2002, p. 256B).

The neural details of the chunking mechanisms are still to be clarified. But the principle as such suggests the following idea for syntax: Instead of the simple principle of keeping the order of the sequence of numbers, consider that the sequence consists of properly ordered word forms of a sentence. Instead of your knowledge about structure relations of numbers your knowledge of language structure organization becomes relevant. It provides the attention concentration on background knowledge for a syntax consisting of word categories and syntactic combination rules. Accessing in parallel these neural organizations your pre-frontal brain's attention might be able to build the syntactic hierarchy.

Such a procedure recalls the middle part of the analysis schema empirically established in Friederici's analyses (2002).[5] Thus the *system of syntax rules could be understood as forming a background system* that can be accessed similarly to a chunking buffer. This idea looks like a kind of complex adaptation of the background knowledge-based original verbal rehearsal process. In this perspective the grammar is simply a background system for *background supported* grouping of hierarchy-based rehearsal activating temporal sequences groups, groups of groups etc. More detailed and technical analyses of structures and procedures will be discussed in Chapter 8.

[5] See also B.J. Baars' (2007b) overview of Friederici's systematic on page 338.

4

Combination and integration of intelligent thought and feeling

4.1 The phenomenon of creativity and advanced forms of experience

The previous chapters concentrated on neural perspectives of concrete semantics such as seeing, acting, memorizing, and meaningful understanding. But in general the frameworks discussed were automatic processes in perception–action hierarchies. Their initiation might be caused by a conscious intention but their detailed execution does not involve or even require conscious control and intentional direction. Even grammatically correct speech acts organized in Broca's and Wernicke's areas have been characterized as automatic neural processes. The previous chapter has explained the surprising fact that, though automatic, the organization of vision requires a much more complicated system of cooperating brain components. They are most appropriately explained by the procedures of saccadic eye-movements when they organize identifications of objects and situations.

But not all organizations of vision and speech are automatic. The organization is essentially different in creative acts of human vision and thinking, and also in concrete feeling of self, other self and social cooperation. The present chapter will study these phenomenological aspects and neurocognitive organizations that are involved. Given the phenomenal complexity I will not begin with neurocognitive models but rather with phenomenological considerations of vision and visual imaging. The reason is that the status of different stages of complexity is best understood by explaining the steps leading from the simplest acts of concrete vision over standard visual identification of the objects of the environment to experienced studies of works of art. The subsequent sections will show that creativity in understanding of visual art may be given a key role since it has some similarity with other ranges of creativity, as for instance science. Creativity is even relevant in forming personal meaning of self and other self.

The phenomenological background provides the foundation for later sections that concentrate on corresponding neurocognitive models that organize the role of the central neurocognitive components, that is, the combination and integration of intelligent thought and feeling.

4.2 Steps towards creativity of visual thinking

Studying certain simple phenomena of concrete semantics opens new perspectives. Let us again consider some different types of vision studied in the previous chapter. Jackendoff (2002, sect. 10.5, pp. 306–311) discussed the simplest type of vision that occurs when seeing something whose gestalt is unclear and has no known feature of one or another kind. At best one can say something like "There is something; I don't know what!" As perceptual feature there is not more than seeing something moving in an unclear way. There are no other features perceived by the fovea, and no features perceivable outside of the fovea, no typical peri-foveal features that could cause specific saccadic eye movements. We are confronted with a visual act without specific saccadic eye movements! Jackendoff says that the something presents at best the *momentary indexical feature* of visual location, brought linguistically in *relation to a deictic particle*: "There!"

The next stage of complexity is one in which the perceived features of the fovea provide already distinctive visual features that allow us at least to identify a simple gestalt or a feature characteristic that allows us to say "There's *something like* an x" where "x" may perhaps name an object or an animal. This is a *local identification act.* The next complexity stage of saccadic identification was exemplified by the fox identification (Figure 3.2) or the identifications of objects on a table (Figure 3.3) in the previous chapter. These saccadic object identification acts are *object identifications based on saccadic focusing* sequences. The objects that can be named, as for instance the fox, the plates or the set of cutlery, are identified in appropriate cortical processes identifying their gestalt, their name or both. These processes are usually automatic and do not need control by consciousness.

This situation may also happen in a standard look at a picture. Consider again Figure 3.3 but neglect the saccadic movement features such as the lines or the circles. In this case you may apply a quick look at the picture that allows you to say that on the picture there are plates, a cutlery set, a glass, a mug and some other things. This way of just naming identifiable objects on a picture is not the manner that is characteristic for an artistic interpretation of a piece of art.

Different experts of art give different interpretations in the details. But they all agree that much more is involved in looking at a painting or an aesthetic photo than just naming the things that can be seen. In his book "*Visual Thinking*" (1969) Arnheim explains this type of thinking as follows:

> *It is in works of art, for example, in paintings, that one can observe how the sense of vision uses its power of organization to the utmost. When an artist chooses a given site for one of his landscapes he not only selects and rearranges what he finds in nature; he must reorganize the whole visible matter to fit an order discovered, invented, purified by him. And just as the invention and elaboration of such an image is a long and often toilsome process, so the perceiving of a work of art is not accomplished suddenly. More typically the observer starts from somewhere, tries to orient himself as to the main skeleton of the work, looks for the accents, experiments with the tentative framework in order to see whether it fits the total content, and so on. When the exploration is successful, the work is seen to repose comfortably in a congenial structure, which illuminates the work's meaning to the observer.*
>
> *More clearly than in any other use of the eyes the wrestling with a work of art reveals how active a task of shape-building is involved in what goes by the simple names of "seeing" or "looking.* (Arnheim 1969, pp. 35–36)

Two years later (1971) Arnheim gives an even more differentiated characteristic of the visual artist. He now writes:

> *It is necessary [for the artist] to distinguish between the balancing of [art relevant] forces in the perceptual field itself and the "outside" control exerted by the artist's motives, plans and preferences. He can be said to impose his structural theme upon the perceptual organization. Only if the shaping of aesthetic objects is viewed as a part of the larger process, namely, the artists coping with the tasks of life by means of creating his works, can the whole of artistic activity be described as an instance of self-regulation.* (Arnheim 1971, p. 34)

Since knowledge of the principles of aesthetic vision is not very familiar it may be of interest to further explain differences of the aesthetic "inside," that is the balancing of art relevant forces in the perceptual field, and the mental "outside" of intentionally underlying motives, plans and preferences. Concerning the former the difference between paintings and art-photos is particularly instructive. Marlene Schnelle-Schneyder (1990, pp. 32–37) explains that there are important differences between painting and photos of aesthetic intention. For painting the original frame for the composition of element forces is the empty surface and its limits. In the act of the painter, each form gets its particular position on the surface and in its relation to the relative locations of the other forms. Ultimately all forms compose a system of aesthetic composition expressing the distribution of the aesthetic forces. This artistic construction differs from the one that characterizes photography. In this case it is the position to the environment that the photographer selects. This position decides in which way the optical arrangement of environmental elements is related to the camera's input. Contrary to painting the original character is the camera-determined perceptual visual field. The photographer cannot compose intended components. The configurations of perceived parts are given. But their field of relations can be modified by changing the positions and orientations of the camera relative to the given environment. This does not change Arnheim's criteria of the balancing of aesthetic forces. But the difference is how the painter and the photographer *select* the aesthetically most appropriate relations. Schnelle-Schneyder explains the complexity that emerges when different experience features must be taken into account, such as the tensions of movement and rest, colour and mood, colour and emphasis and many other forces (Schnelle-Schneyder 2003, pp. 194, 260, 273). It must however be acknowledged that the situation is more complex in staged photography and analogues and digital montages.

In any case the complexity of visual thought that is required in production and perception of art shows on two different levels, the arrangement of aesthetic forces of complex vision, and its "outside," the selection of perception transcending mental control elements exerted by the artist's motives, plans and preferences.

Arnheim (1971) exemplifies the difference in interpreting the sculpture of a Gothic Madonna: He first concentrates on the *structural theme*:

> *One notices a lateral deviation from the fundamental frontal symmetry of the standing figure. The Virgin is deflected sideways towards the secondary centre of the composition providing a support for the child. The deflection is "measured" visually by the spatial orientation of the scepter which is tilted away from the vertical line like the needle of a compass. Here then is the basic theme: the interaction between the majestic symmetry, verticality, and completeness of a queen and the small but potentially powerful child that has sprung from her and receives its support from her. The relation of mother and child allows, of course, for innumerable interpretations, differing in the distribution of weights, of activity and passivity, dominance and submission, connection and segregation.*

Next comes the *mental interpretation* "outside" of the visual variation:

> *The sculpture presents the human relation between mother and child in general; the theological relation between Virgin and Christ in particular; the fact of being conform to the formal and expressive requirements of a particular style; and a particular artist.*

4.3 Creativity in the perspective of scientists

I will now turn to phenomenological studies of scientists that open the perspective of a new variety of creativity.

Among the most interesting arguments are those of Einstein and of Penrose. Both know quite well the positive properties of mathematical statements based on principles in order to clearly mark the necessity of crossing limits of thought and of developing fruitful competences of creative thought. These may lead to extended consciousness beyond some types of limitations of formalisms.

Penrose presented the following lists of *typical varieties of procedure* that either do or do not require extended consciousness (p. 411).

Access of extended consciousness needed:	Consciousness not needed:
'Common sense evaluation',	'Automatic organization'
'Judgement of truth',	'Following rules mindlessly'
'Understanding',	'Pre-programmed action types'
'Artistic appraisal'	'Algorithmic operation'

Based on his experience as a mathematician Penrose (1989, p. 411–13) claims that the hallmark of creative consciousness is a non-algorithmic forming of judgements. He presents the following simple example: "One will have learned the algorithmic rules for multiplying two numbers together and also for dividing one number by another but how does one know whether, for the problem in hand, one should have multiplied or divided the numbers? For that, one needs to *think*, and make a conscious judgement. ... From time to time, one may need to check that the algorithm has not been sidetracked in some (perhaps subtle) way."

It is often thought that verbalization, or the use of formulas and their imagination, is necessary for mathematical thought. This view is falsified by the reports of excellent mathematicians or theoretical physicists. One of them was Albert Einstein. He writes in a letter:

> *The words or the language as they are written or spoken, do not seem to play any role in my mechanism of thought. The psychical entities which seem to serve as elements of thought are certain signs and more or less clear images which can be "voluntarily" reproduced and combined ... These above mentioned elements are, in my case, of visual and some muscular type. Conventional words or other signs have to be sought for laboriously only in a second stage, when the mentioned associative play is sufficiently established and can be reproduced at will.*[1]

[1] Cited from a letter that Einstein wrote to Hadamard in R. Penrose (1989) p. 423. See also Penrose's statements cited above.

These remarks relativize the rather direct relation that theoretical linguists and logicians often assume to exist between language form and meaning. The relation may be typical for common sense utterances or thoughts but not in the range of creative thinking. The previous section explained the phenomenological details in production and interpretation of art. Scientists often tend to think that the situation is radically different in science. Experiences of certain mathematicians, as for instance Hadamard and Polya and scientists like Einstein and Penrose are usually neglected.

In this case it is interesting to add directly the phenomenological ideas of the neuroscientist Fuster. His detailed interpretations in the framework of his neurocognitive models will be discussed in subsequent sections concerned with neurocognition of creativity and other processes that differ basically from the automatic processes of the perception–action cycle explained in Chapter 2.

For Fuster as well as for other scientists thinking about creativity in science is more natural and encouraging than discussing creativity of art, even when they agree that creativity also exists in scientific thinking, though it is often rather assigned to intelligent thought. Creative intelligence is characterized by the ability to invent goals, projects, and plans or to discover principles underlying the reality. Both develop from a broad base of knowledge, implicit and explicit, that was acquired in the past by automatic attention and perception whose memories were more and more symbolized by language in childhood and adolescence, often accessible as noted and symbolised in language. In adulthood a vivid background of common sense and folk psychology is a rich framework of background knowledge which developing capacity of creative intelligence accesses, re-interprets and re-organizes, adapting them to fruitful and efficient knowledge and integrates them into frameworks of life and reality. This is a brief account of basic phenomenological features. They can be enriched by some further characteristics. It seems that the dominant procedure of creative intelligence is not convergent thinking, that is, inductive and deductive reasoning, which converge toward logical inferences and the solution of problems. More typically it is *divergent thinking* free of logical constraints, autonomous and to some extent free-floating, reliant on imagination and minimally anchored in immediate reality. Creative pieces of knowledge, that is, cognits, emerge mainly from *divergent thinking* (Fuster 2003, pp. 242–243, 245–246). In Fuster's words we may say that, in the ranges of reason, to create is to make new cognits out of old ones. In recalling Einstein's report these processes rely on interaction of vivid memory and vivid imagination.

4.4 Some special aspects of the pre-frontal cortex

It must be acknowledged that there are few reliable and directly pertinent studies of brain function in creative intelligence. Thus detailed and established neurocognitive knowledge of the object is very limited. But models that have been developed by experts of neurocognitive studies are very plausible and provide excellent bases for future studies. In general they are based on phenomenological psychology and considerations about language and are thus supported by careful interdisciplinary analyses.

In my view the models of Damasio and of Fuster are particularly fruitful. The essential components relevant for the phenomena discussed so far and others still to be explained have the pre-frontal cortex as the integrative centre. Contrary to the automatic processes of the perception–action cycle and those of the body-based nervous system the pre-frontal cortex accesses the dynamic automatic memories, modifying them in creative thinking or evaluating the emotional status based on body experience. Recall again the motto: "To create is to make new cognits out of old ones".

But before discussing these models of neurocognitive organization I will mention some general aspects. Let me first recall the essential aspects of the perception–action system that was discussed in Chapter 2. It was presented as the automatic perception–action cycle: *Pre-frontal areas* of the cortex organize attention-selected access to patterns of modal and multimodal levels of the perception–action hierarchy. Considering now the pre-frontal cortex we may say that it forms the executive system operating on the hierarchical system. Due to its ability to access practically all hierarchical components the operations of the prefrontal component are called *heterarchical*. In processes of creative thinking and creative intelligence, which we discussed above, the heterarchical accesses operate as follows: They select cognits (neural representations of pieces of knowledge) and form or combine *mental constructions of intelligent and intentional use of simple and standard thought components, thus serving constructive aims of thought, including reasoning, planning and creativity.* In the human species, both the automatic and efficient *exteroceptive organization complexes*, and the *intelligent and intentional system*, became possible due to a radically enlarged pre-frontal cortex. On its basis new kinds of network systems developed: more complex automatic hierarchies and selective higher order optimized heterarchical attention organization. The latter allows a *core system for intelligence*. In combination with communicative development of language and of memory-based thought and reasoning, and its neural connection with the processes of thought, pre-frontal *attention gradually improves the competence of intelligent practice, in particular in conscious exteroceptive behaviour of the human species.*

It is particularly instructive to briefly recapitulate the ontogenetic development of the perception–action system together with the pre-frontal cortex. It is marked by internal goal-directed organization of increasing complexity at progressively higher levels of the cortical hierarchies. At every stage of intellectual development of children higher levels of automatic perception and action are brought into play. This growth involves the formation of a vast array of complex cognit patterns, particularly in the posterior cortex, among them the symbolic cognits of language in Broca's and extended Wernicke's areas of the left cortical hemisphere. It is the functional role of the heterarchically operating pre-frontal components that enables in adolescence and adulthood the formation of intricate behavioural sequences, logical constructs and elaborated sentences, but in particular creative combinations of new imaginations and knowledge combinations. The reason is that, at this age, the mind becomes less stimulus-bound and more the master of the reasoning self. This transition is supported by the development of more advanced and creative forms and meanings of language. Time and deliberation intervene ever more often heterarchically between stimuli and response reactivity and correct automatic usage of common language thereby setting the ground for creative intelligence.

It is also important to mention some surprising aspects that lead to phenomena and processes to be discussed in the following chapters. *Creative processes* do not only result from selective access of the pre-frontal cortex forming new structure combinations in the perception–action hierarchy. There are also *inputs from sub-cortical and mid-brain formations*. Fuster emphasizes that they contribute to the creative process functional inputs from *drive, motivation and attention*. They also have influence on the creative process by indispensable energizing tone from body internal biological sources. Fuster suggested that also transmitter systems of sub-cortical origin are involved. He adds that the limbic system and neocortex send excitatory inputs from the value system that facilitate and maintain the process of creative intelligence (Fuster 1997). Included in those systems are the neural networks that represent a wealth of social, aesthetic and ethical values. These are Fuster's extensions of the core processes of cognition, mainly organized in the cortex, in some respects supported by non-cortical, interoceptive influence. But these indications are still limited. The following chapters will change the perspective. So far the focus was only on the integrated organizations of perception, action and intelligent thought, that means on the core of the common understanding of objective cognition. A new framework will radically change the cortex-based classical understanding of cognition.

4.5 A linguist's critical discussion: an interlude

Given the context of this book the reader might ask at *relevance* the understanding of the neural organization of the pre-frontal cortex and of the nervous system has *for language?* Linguists especially, who have not yet incorporated the implications of the cognit and cognitive network ideas presented in Chapter 2, might be *tempted to ask* how the categories and rules and structures of language form, of phonology, morphology and syntax, are in the brain, and how the elements of semantics, for instance abstract categories and meaning relations, as well as the concrete properties of perceptions and actions, could be represented? They are sceptical that the common linguistic knowledge – *category configurations* and *rules* or other *denotations* – are written in the brain by something that could match formal symbols or letters. And what could be a heterarchical network that selects activities of networks? In other words, they doubt that the brain matches an efficient definition of a formal linguistic system or a programmed computation process. And they are right in their scepticism.

How then are *linguistic* structures and their categories organized in the brain's memory? The first attempt to answer this question would be to explain the *neurocognitive notion of memory*. It will not come as a surprise that somehow the activity states of the brain's *neurons* will have to serve as representation material for categories and the activity-generating neural networks will have to take over the role of generative rules, the task of distinguishing representation positions of category tokens and other organization problems. These aspects were already suggested by the generalization of the Jakobson–Teuber principle. Their unfolding in Chapter 8 might contribute to represent categories, category arrangements and possibly also rule distinctions.

As was already explained in previous chapters, distributed neurocogntive networks represent larger dynamic complexes of knowledge waiting to be activated in appropriate neural states and contexts. It was even discovered that their arrangement forms a hierarchy of dynamic networks – the perception–action cycles of figure 2.3 – whose interactions generate activity structures. I believe that language of more or less advanced structure can indeed be organized in the automatic *perception–action system*. After a learning period even self-embedding structures which are initially supported by internal attention-based selection will finally allso become automatically organized. The situation is probably different for structure-dependent formal semantics that in many cases involves pattern organizations of reasoning. Here, *selective heterarchical access* to different structures, stabilized for a short term, might require the *participation of the attention system* organized in the pre-frontal cortex. In any case the relation of language expression form and meaning representation requires connections of *distributed activity of distant brain areas.* They are represented by *specific and particular multi-modal connectivity.* Learning the frequently occurring relations of word-form and meaning lead to experience-based fine-tuning of cortical networks, thus stabilizing automatically generating processing networks. Ultimately the continuous regular use does no longer require attention-based heterarchical influence by the pre-frontal cortex. The situation may be different for rare words.

In contrast to the automatized characteristics of perception–action–thought combination we would expect that concrete semantic aspects of visual objects and situations depend on multi-modal mutually interacting sound–vision pattern connections. More advanced abstract constructs of reason may indeed be selectively activated by influence of the pre-frontal areas, possibly combined with closely related support of sub-cortical nuclei. This system of combined perception–action and objective thought processing applies to facts of the external world. Many aspects have already been presented and analysed in previous chapters.

When however, imaginative thought is required specific influence of pre-frontal areas select intelligently and temporally in the appropriate moment relevant structures that are memory and knowledge accessible in perception–action areas and in cortical areas.

The linguist who recalls that Broca's and Wernicke's areas are parts of these cortical systems may agree with some plausibility of the system, even with distinctions of simpler and more advanced organization forms for language production and understanding. But *linguists still may have problems* concerning the appropriate and instructive representation. They might say that, in principle, all of what has been said about the perspective of neurocognition may be alright, but that, from their familiar formalist or imagist representation perspective, there is doubt whether current linguistic knowledge can be illustrated by corresponding neurocognitive examples for descriptive details of morphology, or syntax or semantics etc. Here, neurocognitive theory will have to pass. It is still impossible to present pieces of syntactic knowledge as sections of the perception–action cycle or, using a more general term, as a part of a neural syntax memory. Chapter 8 will solve the problem by proposing and discussing some *working model* ideas. For the time being, the linguist is invited to believe that, some day, detailed neural representations compatible with abstract formalist rule systems will be possible. In the meantime, he should participate in the study of arrangements and process interactions of mind/brain components organizing meaning in the wide sense.

4.6 Introduction to the integrated mind/brain/body organization

As the previous chapters have shown Fuster's publications provide excellent explanations of cognit systems, cognitive networks, brain architectures on different scales and connection-determined time organizations of binding. Together they present a systematic and "objective" description of cortex-based perception, action and thought, a scientific core for understanding classical cognition. The organization of language form, including aspects of formal semantics located in left hemispheric Wernicke's and Broca's areas, provides an expression system for human cognition, thought and communication. On the other hand all sorts of concrete semantics and pragmatics as well as mental images and imaginations are not organized in these left hemispheric areas of L-cognits. Instead the neurocognitive networks of these cognits are connected with distant areas processing non-linguistic visual, auditory and haptic cognits. Depending on their types they are located at different cortical areas external to Wernicke's and Broca's areas. In principle there is practically no area that does not process one or the other semantic or pragmatic type. Exceptions may be those areas that represent non common-sense notions or meanings, for instance almost all systems of intelligence that are developed in addition to form and meaning of speech, constructed aesthetic and intelligent knowledge systems. The discussions of Einstein's thoughts already provided some examples for the latter.

Still there are some important phenomena of linguistic semantics whose organization requires further components. The cortex-based explanations of *seeing, touching, pointing, grasping, walking, running*, as well as *reasoning, problem solving and intelligent creativity* sufficient. The organizations of *emotion, feeling, types of conscious action, own self, evaluation of social behaviour, personal each-other relations and personal body experiences* are not. The organization of *their semantics has its own status*. Organizing needs a completely new range of mental and neural functions, structures and dynamics, in fact a separate system. One may suspect that the "*objective*" *domain* of the perception–action–thought systems must be complemented by a brain analysis of the "*subjective*" *domain* of body-internal experiences.

Section 1.5 of Chapter 1 exemplified already that the concrete semantics of a sentence like "I feel cold" is not represented in the cortex but rather in the autonomous nervous system (ANS). This important part of the nervous system is typically neglected because the topics of the dominant studies, cognition and thought, concentrate on the brain's cortex. Instead it is the autonomic nervous system that is primarily responsible for what Cannon (1939) referred to as the "wisdom of the body". The system maintains the homeostasis and overall health of the body and influences emotion, feelings and the status of self. With its sophisticated motor repertoire of internal and external movement complexes and reflexes the ANS works continuously to adjust and defend the body's physiology. The importance of these processes was summarized succinctly by Nauta and Feirtag who wrote: "Life depends on the innervation of the viscera; in a way all the rest is biological luxury."

Via spinal cord and ganglia the viscera are kept in contact with the hypothalamus, a central unit of the midbrain that is often called the "head-ganglion" of the autonomic nervous system. Whatever happens the ANS is indirectly con-

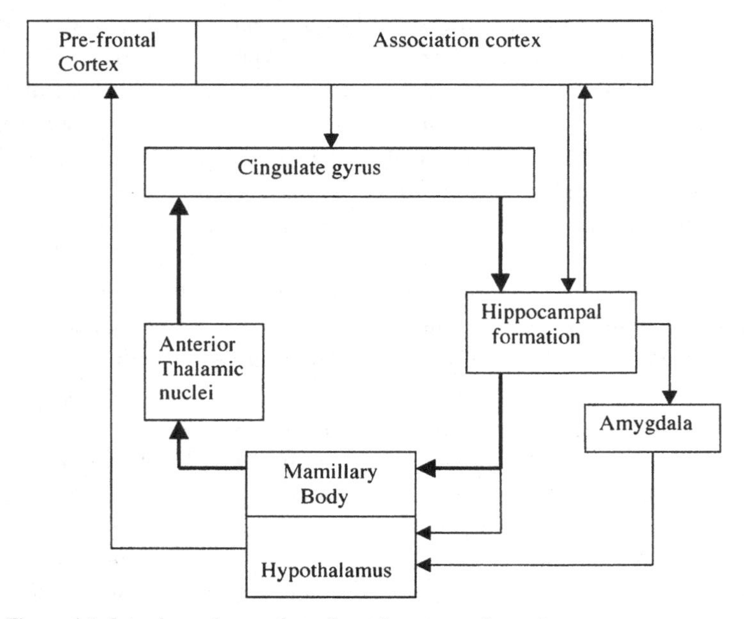

Figure 4.1. Interdependency of pre-frontal cortex and emotion centres.

nected with the pre-frontal cortex via other midbrain components like the thalamus, the hippocampal formation, the amygdala and even the cingulate gyrus, all of them connected with the pre-frontal cortex and the association cortex as represented in Figure 4.1.

4.7 Damasio's development of a radically new approach

In his very fundamental studies Damasio corrected the one-sided studies of cognition restricted to "objective" cognition. He proved the relevance of the body and the internal body organization for the semantics and pragmatics of *feeling* as well as for the understanding of our experiences and knowledge of *self.* During the past two decades he developed new perspectives, theories and explanations. They opened a new framework for understanding a more comprehensive notion of cognition and cognitive neuroscience. The next chapters will show that they also imply a new foundation for the domains of concrete linguistic semantics. Linguistic studies of feeling and self-relevant notions are rare and in their present form not very helpful for bridging the conceptual gaps to possible neural structure organizations in the brain.[2] Before telling Damasio's story I will summarize some basic features.

In combining philosophical and scientific reflection Damasio presented a well-founded network of interesting knowledge about the human being and emphasized that it was a *traditional error* to consider human feeling and the dynamic experience of the inner self as a merely unstructured quantity, as a psychological phenomenon that must potentially irritate our traditional understanding of cognitive and rational mind. Even in modern studies of emotions the neuroscience framework is still misleading due to the limited concentration on "*strong emotions*".

The studies were based on classical conditioning and preferred studies of more extreme states of emotions in mammals showing marked activity in brain areas. Primary theories of emotions were derived from such measurements and distinguish four types of emotional systems in mammals: *Fear,*

[2] This concerns the clarification of the linguistic use of "objectivity" and "subjectivity" to be discussed in Chapter 6, and also the studies about Emotions – for instance A. Wierzbicka (1999) *Emotions across Languages and Cultures: Diversity and Universals*. Cambridge : Cambridge University Press. Her linguistic studies are excellent and her systematic presentation of meanings is useful for comparative studies of the differences in different cultural domains. But constructing a bridge that might lead from here to the psychology of an individual's experience or an individual's brain organization is difficult.

Panic, Seeking and Rage, as prototypical behaviour in animals. In addition to these, three other socially relevant systems have been distinguished: *sexual lust*, the adult mammal's *care-giving system* and the young mammal's *interactive play system*. Other studies were based on observations of *human faces* and *on characteristic body part movements* and their appearances of expressing surprise, happiness, anger, fear, disgust and sadness. Generally all of these experiments relate to forms of behaviour, their interpretation of behavioural relevance (expressed by common words used in folk-psychology) and to brain architecture and brain imaging. This superficial selection of measurements was similar to the fixation on singular behavioural phenomena of perception or action that in earlier times prevented also the discovery of *mirror neurons*.

Damasio studied instead *background emotions* that are not easily accessible to measurements. He reflected about the possible states of *well-being*, *malaise*, *calm* or *tension*. Prominent background feelings include moreover: *fatigue*, *energy*, *excitement*, *wellness*, *sickness*, *tension*, *relaxation*, *surging*, *dragging*, *stability*, *instability*, *balance*, *imbalance*, *harmony* and *discord*. He was certain that they all derived from organizations of the autonomous nervous system and not primarily from the cortex. For a linguist these terms enumerate quite a spectrum of semantic phenomena of a less-known kind.

Thinking and discussing further about emotion and feeling lead to discovering the reasons for their scientific neglect: It is true that, in summary, most emotions do clearly irritate our rationality and provide a characteristic circumscription of the range of phenomena. There is obviously a cultural bias for thinking about or studying these feelings. Whereas scientists may have problems normal people do feel that there are many organized and practical kinds of concentrated and even feeling-supported thought organizations that often *generate important and constitutive meaningful acts* (Singer and Ricard 2008). Some concentrations of this kind involve the understanding and experience of our individual *self's aims* together with the practical organization of *operative self-concentration*. In relaxed situations, states of well-being may also be clear experiences of self-experience. In normal situations simple forms may often remain sub-conscious. Therefore they are often considered as being without much relevance. But well-being supports the background of a proper organization of perception, action and thought.

This brief introduction to the world of emotions and feelings indicates already that ultimately Damasio's views *lead to new theories and philosophies about the role of self, feeling and consciousness*. The special character of his approach is best understood in following his own autobiographic report. The problems that confronted him are in a sense familiar for a modern linguist, as we will soon remark. Here is his story (Damasio 1999, p. 107):

While studying the research techniques taught in medical school and in neurology training some decades ago Damasio was dissatisfied with the foundational and theoretical perspectives dominant at this time. He asked the "wisest" people around about how we produced the conscious mind, and he always got the same answer, very familiar to us linguists: We know because language makes us know. Creatures without language are limited to their uncognizant existence. Some of the "wise" informants may well have added that it is our knowledge of systematic relational structures between concepts and propositions that determines representations of syntax and meaning, and thereby knowledge in general. Language gives us the requisite remove to look at things from a proper distance, the essential base of our consciousness. Other animals are merely determined by learned behaviour without any consciousness. Since language develops by communication and the words of "I" and "You" are only understood by their role in the communicative situation, since they mark speaker and hearer in the communication situation, we experience a concept of self only, so was the argument, after having learned the communicative role of these pronouns.

Damasio thought that this view was fundamentally wrong. The contribution of language to the mind is, to say the least, astounding, but its contribution to the experience of core consciousness was nowhere to be found in common linguistic expressions or structures (Damasio 1999, pp. 108–109). In any case, the answers of the "wisest men" sounded much too simple to the young Damasio. He was sure that we need a corrective. Here are some of his arguments: We must understand that there is a basic interdependence of the processes of cognitive competence, language and knowledge and the parallel feeling that their applications are not merely schematic applications, abstract formal patterns or simple behavioural events. The cognitive aspect seems to be mainly *organized* by neuronal networks *in the brain's cortex*. The parallel feeling is instead closely related to *our self's experiences*, based *on bodily states*. This integration of our personal dynamic of cortical knowledge states and body experience states is even involved in our actions of creativity and the open perspective of infinity in various respects (Fuster 2003, p. 246). Understanding personal experience, language and music, knowledge and creativity in this dynamic context transcends the analytic focus on formal structure constraints. Formal structure constraints are useful in rigorous constructions of proofs, but they block the perspective of the speakers' and hearers' inner experience in their living dynamic as it interacts with knowledge.

For most of the "wise" people Damasio's aim to come at least to an understanding of another notion, namely emotion, sounded even more problematic. To Damasio there were even curious parallels to the scientific neglect. He wondered why in all of these "wise" views there was a lack of perspective in scientific study of mind and brain, namely the *evolutionary perspective*, the study of *homeostasis*, a key to the biology of consciousness, and the noticeable absence of the notion of *organism* in cognitive science and neuroscience.

Fortunately, in recent years both neuroscience and cognitive neuroscience have finally endorsed the study of more appropriate interpretations, for instance Damasio (1994) and Le Doux (1996).

In his books Damasio correctly criticized the situation: The notion of mental competence cannot be restricted to behaviour or limited notions of cognition. The reason for twentieth century's neglect of scientific studies of emotion and feeling[3] was the critical attitude of scientific research that was basically influenced by logical and mathematical progress of formalist analyses. Its formidable development since the beginning of the past century was of decisive influence also for theoretical linguistics, information theory and computer science, and indirectly for cognitive science.

Let us return to the interest of the present book. We want to attain an interdisciplinary account of language that comprises areas that were so far rather neglected. The best way will be to generalize Damasio's and Le Doux's arguments by a critique of formalist restrictions of the studies of language and cognition that do not account for the linguistic experiences of personal and body-based experience. It is true that, so far, scientific studies have mainly concentrated on formally and objectively motivated and systematically restricted formats of languages, assuming that features of life are merely external and secondary phenomena of an underlying formal foundation presented by classificatory and relational symbol structure. Both studies, those of cognition and those of language structure, lost essential perspectives since they were restricted to formally structured expression systems and to logically based meaning analyses, as the "wise men's arguments" confirmed. More specifically, their formalist perspectives reduce content analyses to notational configuration systems, to abstract descriptive units called entities. In the manner of formalists they thought that a formal theory implicitly defines the complete system of essentials in terms of a constructivist world-view of things and events. This view

[3] A more detailed discussion of language frames of emotion words does now exist. It demonstrates the cultural differences involved. A. Wierzbicka 1991, p. 53–54.

claims that even speaking and thinking selves are *basically understood* as classifiable sub-sets of the comprehensive world of things and events.[4]

This was the situation at the time of Damasio's early reflections. But he knew that proving the error of his "wise men" would require empirically and theoretically justified results. He himself would have to present more specific and fruitful arguments and theories based on his neurological and clinical knowledge. He thought that the following reflection gave hints to a better start: In most cases consciousness is the content of *verbal expressions of ongoing mental processes*. If language expressions – that is, word and sentences – are representations of something else, this something else might exist as non-linguistic mental images, realized in cognitive networks. Language expressions might thus represent directly the mental images of entities, such as events, relationships and inference, and thus indirectly identify the very entities. Is it not possible that language use "operates for" *mental-self* processes and for mental consciousness states in the same way that its use operates for everything else? The argument could be generalized: Some words and expressions might indirectly symbolize *self-representing processes and consciousness states.* Spontaneously there are body internal processes in non-verbal form. Only subsequently they are expressed by words, sentences and other expressions. If so, there *must be a non-verbal self* and a *non-verbal knowing* experienced internally by human individuals. The underlying processes mark core features that are components of meanings expressed by the words *"I" or "me"* or the phrase *"I know"*, features that are *completely independent of the deictic role features*, which, strangely enough, could only be effective in a communication situation. Consequently, Damasio believed that it is legitimate to take the phrase "*I know*" and deduce from it the presence of a non-verbal image of knowing that is centred in an experiencing self and precedes and motivates that verbal phrase "*I know*" in an act of thought. Thus emotion, feeling, the momentary experience of self- and other-self seemed to be interdependent.

It is obvious that these thoughts are near to Descartes' famous statement "cogito ergo sum." Damasio thought that this statement implies an error. Here are his words: "Long before the dawn of humanity, beings were beings. At some

[4] Here are some brief indications of the historical development: Fight against psychology (a) Frege (b) Hilbert's geometry, (c) Carnap's structure analyses and the influence of his Syntax on language analysis and Chomsky's re-adaptation of the early 1930's formal symbolic analyses. See status in presentations in Part II . Parallel to anti-network analyses by Lashley in Neuroscience: Consider in Kandel et al. (1995) p. 15a. Lashley was deeply sceptical of the cortical sub-divisions determined by the cytoarchitectonic approaches and distributed analyses. This was subsequently a clear influence on Chomsky's new linguistic scepticism against a possible relevance of neuroscience for linguistics.

point in evolution, an elementary consciousness began. With that elementary consciousness came a simple mind; with greater complexity of mind came the possibility of thinking and, even later, of using language to communicate and organize thinking better. In this understanding, being was in the beginning and only later was it thinking. And for us now, as we come into the world and develop, we still begin with being, and only later do we think. We are, and then we think, and we think only in as much as we are, since thinking is indeed caused by the structures and operations of being.

Without going into further details we should at least reconsider Damasio's fundamental stance presented in his book about *Descartes Error:* "The comprehensive understanding of the human mind requires an organismic perspective" (Damasio 1994). Ten years later he published the book *Looking for Spinoza* (2003) in which his position was based on organism-based arguments of Spinoza's philosophy. Concerning the organism-directed perspective Spinoza's position is in many respects similar to the relevant parts of Leibniz' *Monadology* presented above in Chapter 2, section 1.

Damasio's outline of Spinoza's statements confirms this view: "Mind and body are parallel and mutually correlated processes, mimicking each other at every crossroad, as two faces of the same thing. Deep inside these parallel phenomena there is a mechanism for representing body events in the mind. In spite of the equal footing of mind and body, as far as they are manifest to the percipient, there is an asymmetry in the mechanism underlying the phenomena. The body shapes the mind's contents more than the mind shapes the body's, although mind processes are mirrored in body processes to a considerable extent.[5] Let me add that Leibniz' position embeds the organism-based framework in a mathematically understood infinite framework of the universal reality. The latter contains the fundamental principles of universal harmony. The principled infinite character of each individual is the reason why the individual in his act of organism-based perception and thought has only a reduced accessibility to the infinite details of the universe, and even to the details of his personal constitution. The restrictions of her or his perception do not allow the percipients complete understanding of the universe or of details of the underlying principles.[6] I shall not try further clarifications characterzing the differences of the organismically oriented approaches of Spinoza, Leibniz and Damasio. It is time to return to further discussions of Damasio's details.

[5] A. Damasio (2003) p. 217, interpreting Spinoza.

[6] Cp. reference to K. Gödel's notion of "trans-computational" capacities in humans, p. 354–355 in H. Schnelle (2004).

4.8 Is the notion of self a feature of the first person pronoun?

Most linguists might be sceptical about Damasio's claim that a non-verbal perception or image of the infant may and often does precede learning the use of *first and second person pronouns*. They would think that the linguistic view of the priority of deictic pronominal is absolutely justified. Isn't it true that the communicative use of these words is learned not earlier than during the third year of life, definitely in contexts of communicative use? They will accept the fact that this stage is prepared by another stage in which the possessive particles "my" or "mine" accompany gesture claims of possession, and thus precede the deictic use of "I". Thus, our linguist would insist on the deictic interpretation of the speech act: The spoken words of a sentence utterance mark a momentary time interval and the words "my", "mine" and "I" occur as pointers to the person who utters this sentence at the marked time interval. The linguist insists on the standard meaning of the words "I" or "me," contradicting the assumption that other interoceptive experiences of infants have established meaning features that precede the use of pronoun expressions.

The linguist is correct, as long as he describes one aspect of the meaning, the one that is restricted to communicating information about external facts of the world, namely utterances. Still, the linguist argument is too restricted. Consider the statement "I am feeling bad. I have terrible stomach ache." Here it is true that, in one sense, the speaker makes a statement about himself, more precisely about what *his self feels*. The deictic feature of the first person pronoun is only secondary. Now consider a crying infant. According to mother's experience she knows that the baby feels a stomach ache and obviously feels bad. The baby does not communicate by using a learned convention that the sound of crying refers to feeling nor that it expresses a statement. Thus the infant's crying does not have the *intention* to express the deictic fact that *it* is the cry-uttering individual. The communicative content as such, that is the meaning, is similar to the adult's statement. In the situation of the adult the essential information is also the *expression of the internal visceral feeling of the uttering person*. In this situation of obvious feeling the deictic role of the first personal pronoun is *absolutely secondary* compared with the reference to the personal feeling. In the baby and the adult the emotion and feeling experienced by the self are similar. *Hence both have a self as a base of feeling experience*. This self is an experiential continuity that *later becomes* a meaning component of the first person pronoun, having experiential priority relative to the deictic feature. There are many situation types in which the former remains more dominant than the deictic meaning.

We may conclude that *infants have indeed momentary self-feelings* long before learning the reduced communicative elements of the language, such as the deictic role of "I" and "You". Similar arguments would show that many expressed word meanings, which are acquired during the second and third year, rely on experiences and experience frameworks of the first year. In the meantime their experiential understanding does not need support of conventionally acquired word sounds. More generally: If language expressions – that is, words or sentences – are representations of something else, this something else often exists as non-linguistic mental image, realized in cognitive networks, *possibly independent of word-denotations*. During early childhood acquisitions of words might use directly these internal experience-based mental images. In these cases the language forms and formal meaning relations are not prior to concrete meanings; it is the other way round: Concrete meaning experiences in the inner self provide very often a foundation for developments of language usage, particularly in the developing mind of the young individual. Thus, Damasio's idea is justified. There are many knowledge aspects that infants develop during the first year of life. The next section will not present further details. Here the main point was to introduce the relevance and justification of the notion of *self* by indicating its early availability during early childhood.

4.9 Cognitive- and body-based neural systems and their roles in infants' learning phases

The previous discussions of Damasio's arguments were mental. They did not yet refer to functional neuroanalysis or to neural architecture though indicating some underlying principles, such as the notion of *a* self as a base of *feeling experience* that in turn has foundations *in internal body processes* and *emotion signals*. These indications contrast with core phenomena of cognition and mind, the focus of Fuster's perception–action–thought frameworks that were mainly discussed in earlier chapters.

It is now time to recall some basic remarks made in Chapter 1, section 5, about the basic constitution of the complete nervous system, some of which were also repeated above. It comprises *two basic functions* that are organized by two interconnected global parts of the neural system. Let me briefly summarize some aspects. The two basic function-complexes are *perception–action–thought, comprising language and conception support* on the one hand, and the *self's emotion–feeling* on the other. The former is based on the brain's cortex and the sub-cortical limbic system, together with the *somatic division* of the nervous system. The somatic division provides the central nervous system with sensory information about muscle and limb position and about the environment outside the body. It also sends signals to these areas, thus acting on body movements of the limbs, *breathing*, articulation organs etc. and surface characteristics like the extended feeling of the *skin and touch*. The *brain's cortex* processes the signals received from or sent to the somatic division and generates and organizes the cognitive networks representing cognit configurations of pieces and structures of knowledge about the experiences and theories of the environment. The experience system of self–emotion–feeling is mainly based on the so-called autonomic system of the body.[7] It organizes – in the sympathetic sub-component – the response of the body to stress, and acts – in the parasympathetic sub-component – to conserve the body's resources and restore the equilibrium of the resting state. All three, the cortical system of cognition, the somatic system and the autonomic system, contribute to the body's homeostasis and constitute feelings of self-presence. They appear anatomically separate but are functionally *and centrally interconnected*, thus *combining* the representation of *cognition* with the representation of its *cognizer* in an integrated realization. Areas of the pre-frontal cortex mainly organize the important central interconnection of both components.

It is true that the network details and the efficiency of distant connections and the advanced forms of the pre-frontal system's combination require years and are developed during childhood and adolescence. Their completion is only attained in adulthood, though the more primitive and basic and functionally important interconnections are already available in infants. Studying the early developments is particularly fruitful in the context of our comprehensive nervous system processing. They clearly involve simultaneously the cortical perception–action system as well as the body-based nervous system foundation of emotions and feeling.

Recall some earlier remarks about the developmental stages: During the early months of the infant the modalities of the auditory, visual, and haptic systems first develop on the *phyletic* and the *mono-modal* levels of the perception–action cycle. The different modalities are still isolated. But in the more comprehensive perspective, which also takes care of the autonomic nervous system, additional organizations of experience are involved. Here the visual modality is closely related with parts of the mid-brain, such as the superior colliculi, providing *eye direction and stable focus*. The system of manual touch is complemented with the *skin* organization of limbs and body surface feeling. The articulation and audition system is related first to the muscles of the abdominal, costal and glottal

[7] The constitutions and the processes of the sub-components of the body-related nervous system are described in E.R. Kandel et al. (1995 p.599).

articulation components. The latter also mark the *inferior unit of articulation*, which cooperates with the laryngal, pharyngeal, oral, velar and nasal systems in communicative language utterance. In the case of the auditory system midbrain units, such as the inferior colliculi, operate nearby to the superior colliculi supporting vision orientation. The infant and children's brains must develop the functionally appropriate cooperations and integrations of these sub-systems. It is plausible that the functioning of first mentioned sub-units and sub-areas precede the development of their connections with the cortex.[8]

More or less, the phyletic organizations derive genetically in correspondence to mammal brain architectures. For some functions there is already prenatal operation. Soon after birth the system starts with distinct operations of pre-structured sub-sections of the nerve system and of phyletic interactive organization of primary cortical areas. Their primitive adaptation to environmental conditions will be gradually extended during the second half of the first year. During the last months of the year haptic and visual categorizing become correlated in multi-modal connection, supported by grasping actions, and activated in quasi-synchronized binding processes.

It may be interesting to learn about the studies that provide empirical information about *developmental stages*. These studies rely on two different procedures: extended and comparative observation of competence development, and experimental measurements of situation-specific reactivity in behaviour. I will present some of them, separately for *sound, vision and touch*.

Among the earliest perceptions of an embryo during the last month in the womb are *hearing processes* of the *sounds and rhythms* of mother's *speech* (Karmiloff and Karmiloff-Smith 2002. p. 12). Immediately after birth hearing mother's whispered speech is thus a *familiar experience* for the infant. During the first 4 months the infant starts to gurgle and coo. They utter vowel sounds such as "oooh" and "aah", sometimes even "experimentally" narrowing the larynx leading to husky vowels. Typically their utterances are spontaneously and incessantly produced when emotionally calm.[9] It seems that the infants *like* producing and hearing their own sounds, just as they like their mother's calm sounds without intending already to properly adapt the adults' sounds. At 4–6 months, babies may start to babble (adding consonants "gaga", "dada"). At 6–12 months babies typically babble and enjoy vocal play, as they experiment with a range of sounds. At this stage, brain organization of primary cortical areas may *adapt* distinctive operations for the consonants used in their environment. Much later, when the infants begin to use words and single word combinations the

[8] This evolutionary and developmental character has been demonstrated by R. Rafal (2002).
[9] T. W. Deacon. (1998). The symbolic species: The co-evolution of language and the brain.

communicative play of utterances starts to be combined with attention, intentional and goal-directed speech.

There are thus three different frameworks of motivation, (a) *body emotion-based* own *experiences of calm* combined with the *feeling of general pleasure and satisfaction* generated by sub-cortical processes, (b) *adaptations* of perceptual distinctions to environmental events of spoken sound patterns, and (c) intentional integration of sound utterances to situation-based usage and goal-directed actions. In other words we distinguish types of body-based spontaneous experiences satisfying *self-experience, sub-conscious adaptations to regularities of common usage* and *spontaneously being satisfied by success in communicative or behavioural play or games.*

The functions are clear but the organizational principles of development and motivation are less transparent. Observers will often *not be sure in their evaluation of the motivation.* This is even truer for measurements since they are primarily interested in *measurement results* of developmental or learning procedures. They are not really concerned with the *organization principles* underlying the *foundational innate neural networks* and *the growth of their fine-tuning* leading to organizations of increasing efficiency and satisfaction in the given environments.

In the case of *vision*, the infant focuses on the mother's face very early. There are positive measurements at 10 minutes after birth. It must be assumed that the face configuration is already genetically imprinted in the infant's orientation map, probably not in the cortex but in the map of the superior colliculi of the midbrain. Perceiving mother's face seems to be a *positively marked feeling* soon after birth. More detailed eye movements in looking for objects develop very quickly as well as eye movements following objects. Also saccadic eye movements develop after some months. Their temporal orientation and rhythm will probably also be experienced first as fun and later as important contribution to motor processes, such as grasping and manipulating objects. At this time vision contributes by object form perception to the motor acts of grasping, whereas *grasping* and *object touch* contribute to gestalt construction. Vision and touch become a combined system in perception–action. Recall the description of saccadic eye movements.

Due to different types of feeling sense units in the *skin*, touch connection felt in the grasping hand is basically different from feeling of the body's surface skin, for instance when being in touch with a loved person's body. It is certain that this positive feeling in the body's surface is very important for the neonate and plays a central role in constituting the *positive affections felt by being held in the arm, rocked and hugged.*

During the second half of the first year certain multi-modal combinations develop, for instance the combination of vision, grasping movements and distinguishing touch determined object forms and surface feelings. From about 10 to 12 months onward *play and games* with infants[10] develop and become increasingly differentiated, in particular when children learn the practical procedures of the game's roles during the next year. Play and games are functions of the child synchronizing the perception–action processes, the bodily feelings and the social processes with intentions in self-activity. It is obvious that during this time the courses for the social roles and for self-assertion are set (Erikson 1950, 1977). Expressed in the brain's perspective the integration of the brain components – the nervous system, the perception–action cycles, the pre-frontal cortex, their interaction with the amygdala and the hippocampal system and finally their intelligent and selective organization in practical thought – are increasingly fine-tuned over the years.

The integrated system and the states of the self and of emotional feeling evaluations develop parallel to perception–action and thought processes, gradually increasing in complexity until adulthood. In many respects, even the earliest forms of feeling play a central role in personal relation development, correlated for instance with seeing the mother's face and hearing the sound and the rhythm of her voice.

I think that these developmental expositions made it quite clear that the interplay of perception–action and thought on the one hand and of the emotion and feeling system on the other have a functionally specific character of human beings. The cortical areas are mainly involved in the organization of cognitive processes, spontaneous perception and action in situations and in communicative speech. But the cognitive system is not enough; if it were, the brain of the human individual would just be an intelligent cognitive system, merely constructively different from a computer system. A human cognitive system is just one of the two basic global systems of a feeling person, a system that organizes knowing and feeling and being *personally involved* at any moment.

4.10 Background self, feeling and constitution

Complementing the ontogenetic observations we may also consider *criteria of evolution*. The interaction of perception–action and emotion–feeling is fundamental. In view to their contribution to the processes of the organism's existence, they interact and have complementary functions. That is their efficient way of cooperating, thus generating the strange state we call life and the strange

[10] Very instructive are the descriptions about early childhood in J. Bruner (1983) and D.N. Stern (2000).

nature of the organisms[11] that drives them to endeavour to *preserve themselves*, come what may, until life is suspended by aging, disease or externally inflicted injury.

In a sense this organization principle is already relevant for cells. But in cells the separation of functions is rather assigned to internal and external molecular complexes. The situation is much more complex for organisms. The nervous system of mammals consists of three global network systems: the *perception–action–thought cycle* already presented according to Fuster's working model, the *emotion-and-feeling cycle* and the *pre-frontal system* in which both overlap.

Before discussing their details in another section I want to specify some necessary principles. Already in the case of the perception–action system I emphasized that we are concerned with functional neuroscience and therefore need the combination of mental and biological perspectives of the simple and complex cognits in Fuster's working model. The other perspective was neurobiological in referring to neural networks as dynamic units, *capable* of acquiring states of activity, whose activity patterns become *energy synchronized* by *binding processes*, thus actually representing momentarily experienced states of the perception–action system.

Damasio is right in pointing out that there is a *similarity to the perception system. Feelings are in some way also perceptions*, though in some respects different. Typical visual perceptions correspond to *external* objects or events whose physical characteristics impinge on our retinas and temporarily modify the patterns of sensory maps in the visual system. Feelings also have an objects or events at the origin of the feeling process. It is even true that the physical characteristics of the objects and events prompt a chain of signals that transit through maps of the object inside the brain. Just as in the case of the visual perception there is a part of the phenomenon that is due to the object, and a part that is due to the internal construction the brain makes of it. But there is something different, though certainly not trivial: In the case of feelings, the objects *and events* at the origin are well *inside the body*. What becomes felt is already inside rather than outside. Feelings may be just as mental as any other perception, but the objects being mapped are not external but rather parts and states internal to the living organism, for instance in components of the visceral system. They activate map configurations in parts of the body's internal nervous system whose signals cause sub-cortical and cortical feeling representations. In the emotion and feeling system the original objects perceived are "*internal objects*", objects of an "*internal world*", in contrast to the cognitive perceptions

[11] Called *conatus* by Hobbes, Spinoza and Leibniz.

mapped in perception–action systems. The latter are usually cognition mapped objects and situations of the *external world*. This even holds for mapped language knowledge systems, where the perception–action system (located in Wernicke's and Broca's areas) determines language form perception and action. Here the "objects" of the external world are sound events.

Let us consider an example: You see a spectacular *seascape*. It is outside of the body. But *in* the perception–action system – locations in the modal areas of the visual cortex and short-term memory – there is a located perception, an *internal* perception map of the *external* seascape. In a next step the nervous system has a direct internal means to respond to the internal object percept *as feelings are going to unfold*. Not merely the perception of colour points is important. The signals are simultaneously transmitted to sub-cortical emotion areas which in turn return signals to the cortex where an *emotionally competent perception* marks a feeling *initiating object*. This is the result of the essential procedure, which might also be called "*the emotion–feeling cycle*." The complex system of interacting components is well schematized in Figure 9.32 of Moscovitch et al. (2007).

More specifically such emotionally competent objects activate emotion centres of the body bringing them into an important attention state by distributing signals to other areas as well as sending feedback to the initiating object. Via *centres of emotion organization*, in particular the amygdala, a sort of reverberative process is engaged that finally marks the initiating object, the percept of the seascape (not the external seascape itself!) as the momentary feeling centre. (If the percept is not very clear an intentional check procedure might confirm that the feeling centre received something in the real world and not a hallucination.) Thus the emotion–feeling cycle provides the processing complex. In Damasio's working model the *organizational complex of emotion and feeling* is the counterpart of the *perception–action cycle* in Fuster's working model.

Our seascape example illustrates an event in which the character of the external perception causes internal emotion and subsequently a state of feeling. The externally focused situation is an emotionally competent object. The example illustrates the combination of visual processing of the perception–action system with the emotion and feeling system. The situation is in principle different in *another example*, the *feeling of well-being* at a wonderful day at the *beach*. Here is Damasio's representation: "Think of lying down in the sand, the late day sun gently warming your skin, the ocean lapping your feet, a rustle of pine needles somewhere behind you, a light summer breeze blowing, 78 degrees F and not a cloud in the sky. Take your time and savour the experience" (Damasio 2003, pp. 83–84). You just are feeling well in a calm situation. In spite of the fact that there are many external elements that might be objects for your perception–action system you do not focus any of these nor do you analyse their combined mapping in order to have a clear picture of the situation. Instead, contrary to most situations, in which the external world is dominant, and the self-feeling is in the background, you are now in a situation in which the present external phenomena are background experiences. They are not "perception focus competent", that is, arousing attention, but they are "emotionally competent"–where the *continued feeling of pleasure and happiness substitutes the attention focus.*

Damasio preferred to study the phenomena of *mild emotions*. They certainly contrast with measurements of *explosive emotions*, preferred by neuroscientists concentrating on experimental measurements. They are typically concerned with attention-focused perception or action without considering possible background feeling, and not at all with the dominant feeling experience that pushes external attention focus into the background. The background feeling of self or of our sensory-motor organs' and cortical networks' activity is usually considered to be merely *subjective*, and is consequently not studied and definitely not considered to be *objectively* present. In fact it is considered to be mysterious. This stance may still be a result of the development of scientific perspectives of the previous centuries. During the nineteenth century there was, for instance, much discussion about the difference of the *objectivity* of experiences and theories of the world and the *subjectivity* of the experiences of the underlying ego or self. Since the latter is not explicitly conscious there is a tendency of considering it necessarily to be *background experience* without any relevance for the truth of external facts.[12] Our example illustrated that this stance is not justified; feeling may be experienced in the foreground, leaving the checking of truth in the background.

Let me add an additional point: In human beings rational thought as it occurs in self-concentrated reasoning or computing may be combined with imagination on the one hand and positive feeling that accompanies the rational thought and enhanced pleasure when the solution is found.[13] These combinations

[12] J. Searle (1992) writes on p. 77: "We have certain background ways of behaving, certain background capacities, and these are constitutive of our relations to the consciousness of other people." I think that his position is similar to Damasio's. In Chapter 6 of the present book I will return to the role of these phenomena in the context of linguistics, particularly in the context of Langacker's (2008) frameworks and his notions of objectivity and subjectivity.

[13] Imagination and positive feeling are even fundamental in mathematical problem solving. See The intuitive words in Polya (1965), when he writes on p. 117–119. "In teaching mathematical problem solving you must know the tricks: Be interested in your subject! Try to read the faces of your students, try to see their expectations and difficulties, put yourself in their place. The art of being a bore consists of telling everything (cf. Voltaire). See Also H. Poincaré Le raisonnement mathematique, a section in "Science et methode" 1908.

are results of biological evolution and do not exist in mechanical constructs of computers. There is much research of computer developments formally simulating as-if emotionality with natural appearance.[14]

And still another point: As just indicated cognition and emotion can cooperate in advanced intelligence. Here their combination relies certainly on appropriate organizations of the pre-frontal cortex. But there are also more primitive and more direct automatic combinations between perception–action and emotion. Damasio described and enumerated some phenomena. Here background emotions continuously underscore the subjects' actions. Gestures and behaviour combine in indicating an observed subject's background emotion. In our linguistic context the following remarks are the more interesting ones: When the observed subject speaks, emotional aspects of the communication are separate from the content of the words and sentences spoken. Words and sentences from the simple "Yes", "No" and "Hello" to "Good Morning" or "Good-Bye" are usually uttered with a *background emotional inflection*. Damasio is correct to state that this inflection is an instance of *prosody*, the musical, tonal accompaniment to the speech sounds that constitute the words. Prosody can express not just background emotions, but specific emotions as well. For instance, you can tell someone, in the most loving tone "Oh! Go away!" and you can also say, "How nice to see you" with a *prosody* that unmistakably registers indifference (Damasio 1994). In all of these cases there is a common "original object level" for expressions, namely the sound events. But the levels of "meaning processing", communicative content expressed by speech sound patterns and accompanying emotions expressed by prosody, have different origins.

4.11 The systematic organization of the three components of the nervous system

Let us return to the basic reflections about neural organizations involving emotion and feeling. The mental characters of the examples – the "*beach example*" and the "*seascape example*" and even the behavioural or linguistic contents and content combinations of expressions – are not really helpful. Our primary question is rather *why we feel the way we do*.

Let us first recall again the *evolutionary perspective*. The organism certainly relies daily on *perceptions and actions* but *vivid engagement* pertains to attention

[14] I. Wachsmuth (2008, pp. 279–295). The principles of my own scepticism are expressed in H. Schnelle (2004).

organization as neurocognitive science says. Our discussions of the saccadic eye movements and production of gestalt perception made it clear. The same kind of operations may become involved in solving a practical problem of a difficult object manipulation. The perception–action description refers to the structure organization of the perception–action networks. But being a person our self feels that there is more. The events were part of our life governance and are simultaneously felt *either fluid or strained.* The fact that we, sentient and sophisticated creatures, call certain feelings positive and other feelings negative is physiologically directly related to the fluidity or strain of the life process. Fluid life states are naturally preferred by our inner dynamic, our *conatus as Hobbes, Spinoza and Leibniz* said. Our natural tendency gravitates toward them. Strained life states are naturally avoided by our *conatus. We stay away.* It is quite obvious that the case of saccadic eye movements and experienced gestalt perception is fluid and our self-feeling is completely shadowed. But in the case of solving a difficult practical task perception and action may be strained. The self is clearly felt as the unit that feels. Feeling also the importance of the problem solving it may support concentrated arousal of attention. There are certainly also cases of life governance that are strained but lacking the self's motivation. In this case we tend to stay away from further activity.

Still, the evolutionary relevance of these internal modes of *conatus* does not really help in answering the question *why we feel the way we do.*

The conatus-based answers in the evolutionary framework are metaphoric. What we want to know is: "How does our brain organize situations that result in constituting our feelings?" In a compact form Damasio emphasizes his guideline: *Feelings are also perceptions.* Certainly there are differences between the *perception–action–thought* system and the perception organized *self's emotion and feeling* system. Two words indicate the original difference: The former represents or applies in specific ways to objects or situations of the *body external* world, the latter represents or applies in specific ways to *body internal* feeling characteristics of *inner objects* or *situations*. But it is also true that often the external world experience and our self-feeling are combined by integrating both processing systems. But still, even here the hints do not yet determine in a more precise way the specifics of the difference.

Perception, action and thought have already been explained in previous chapters. Typical internal objects are the heart, the stomach and other viscera. Whereas perception experience gets its cognit-generating signals ultimately from the light waves rich are sent from the external object, for instance a tree, registered in the retina, transmitted via somatic pathways and causing the activation of the cognits in the perception–action cycle. Internal to the body the stomach's peristaltic may send signals transmitted by the somatic division to the midbrain and lower forebrain comprising also emotion units, such as the amygdala. The momentarily configured patterns of body internal maps are due to the internal construction that the brain makes of it, namely feeling representation, for instance in the case of stomach feeling. The skin exemplifies other types of inner experience. Imagine for instance that your skin signals a mild warm breeze on a wonderful day in a beautiful environment.

In my understanding we must distinguish three global components of the complete neural system

Integrating the *two automatic processing systems* of

– body external information – (the perception–action system)
and of
– body internal information – (the nervous system)
and

selective attention and intention concentrating and integrating processing system

- based on pre-frontally organized combination, the self's intelligence, and on
- the body external and internal processing, the conscious life's self-integration.

The latter rely on the other two, comprising for instance the pre-frontal procedures of vision and advanced forms of speech organization, already

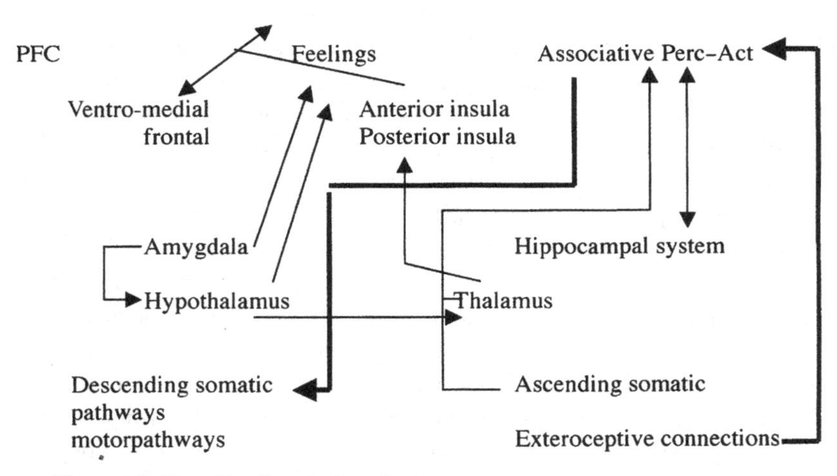

Figure 4.2. Signalling from body to brain.

discussed in the previous chapter, and the ventro-medial pre-frontal region presented, together with the other feeling-involved organizer components in Figure 4.2 (adapted from Damasio (2003), p. 59)

In any case the *three listed systems overlap* and the overlapping units in the pre-frontal cortex contribute to integration. In principle, cognitive and emotional systems are continuously linked in *rich two-way combination*. Certain thoughts evoke emotions and vice versa. Thus, it is not only the case that word or sentence expressions combine form-based meaning content and prosody-based feeling. Due to the brain's organization, perceptual facts, feelings and self-foundational thought orientation are combined and often explicitly integrated, thus contrasting radically with artificial intelligence mechanisms.

PART II

Introducing linguistics to neuroscientists

5
Introducing formal grammar

5.1 Our dynamic perspective

The intention of the first part of this book was to familiarize linguists with an overview of neurocognitive science in the wide sense. This second part looks in the other direction: Linguistics should be introduced to neuroscientists.

For the linguists the first part of this book presented neurocognitive perspectives of brain architecture and brain dynamics. The analyses and descriptions of distributed brain areas of the cortex and sub-cortical areas and nuclei were related to psychological functions supported by the interactive cooperation of brain components. The reality of language form organization in cortical core areas was justified about 150 years ago.[1] Thus language was marked as a central psychological function in the brain. But I also emphasized that, though very important, the organization of linguistic *forms* in phonology, morphology and syntax must be supplemented by organizations of *concrete meanings* of perception and action. Form and meaning cooperation require the interaction of form organization with perception–action and feeling-based brain processing. In fact almost any Broca-Wernicke external area can contribute with its specific organization of concrete semantics by processes of perception and action, attention and intention, memorizing, memory recall of autobiographic and systematic knowledge involving data and processing of knowledge data based on reflective competence systems of conceptual planning organization, emotion and feeling, internal body experiences and self-experiences. Normally these meaning organizers, which are usually distant from Broca's and Wernicke's areas, constitute *universes of concrete meaning and background understanding*.

[1] Broca discovered in 1861 the fact that the reason for certain aphasias was lesions in a specific area of the frontal cortex. Wernicke discovered another kind of aphasia in the posterior cortex. Wernicke proposed an ambitious model of how the brain processes language. Basic aspects of the model are still used today.

Both are actively integrated by "neural network binding," thus constituting the understanding of situation, discourse and thought.

Details of the globally distributed locations of the language-relevant mental functions were usually determined by systematic and careful reflections based on empirical studies of lesions, various forms of brain imaging and microelectrode measurements. These studies led to descriptive and explanatory models of neurocognition. On the microscopic scale details of local neuronal circuits, called neural modules and columns, have been analysed and characterized in their architecture and understood as micro-network components of areas. The exact details of dynamic organization are not yet empirically established. Our understanding of these micro-networks comprising hundreds of single neurons is much less advanced than the functioning of each single neuron, which is determined on the molecular scale.

This introductory text is merely a summary of the chapters of Part I. We now turn to the ideas to be presented in the following chapters of Part II, which are motivated by the challenge of familiarizing neurocognitivists with linguistics. I hope that, after a while, some ideas may suggest possible neurocognitive counterparts to linguistic descriptions, perhaps even general ideas for bridging linguistic observations, theories and models to neurocognitive networks. This is an optimistic view because the situation is very difficult. Obviously neurocognition is a specific discipline that contrasts radically with the basic frameworks of linguistic studies, both on the levels of empirical studies as well as on the levels of empirically plausible working models. Empirically technical measurements and observations of behaviours in tests or after brain lesions contrast with empirical studies in linguistics. Here the linguist concentrates carefully on classes of expressions trying to clarify questions of equivalences or contrasts of forms and meanings. Considerations of working models or theories are also different. The neuroscientist refers to brain architectures and dynamic processes, the linguist to systems of linguistic categories and their systematic order and constitution. Despite these differences empirical cooperation is possible. Linguistically systematic sets of test conditions are often helpful in *defining* the execution of neurocognitive *test procedures.* But in view of inventing and describing *fruitful working models specific experiments* of observation, brain-imaging measurements, sometimes even microelectrode measurements as well as interpretations of behaviour resulting from brain lesions are not enough. They do not provide systematic frameworks of analysis or clarifications of the roles of principles for potentially interdisciplinary correlation of language structure, brain architecture and brain dynamic.

It must be assumed that just the levels of connective brain architecture would be of particular interest in our question of how language form, meaning and usage are organized in the brain. But it is true that linguistic studies of language structure and usage do not yet profit from the recent systematic study results and models of neurocognition. It is understandable that they rarely relate their studies to functional neurocognition models.[2] But it is just on this level of modelling that mutual understanding should be developed. The first chapters of the book explain why I am confident that interdisciplinary studies will develop descriptive techniques that will gradually improve the mutual correlation of neurocognitive modelling with linguistic descriptions.

Here now is the perspective of the following chapters. I shall present modern studies of linguistics that were originally developed in the classificatory tradition of more than two millennia ago. It will become clear that the tradition of core linguistics concentrated on descriptions of form, meaning and context of usage. They led to frameworks of grammar and lexicology as foundations of conceptually precise analyses of grammatical categories, concepts, classifications and configurations as well as to rules of their practical usage in language teaching and to extensive empirical studies comparing phenomena of well-formed language standards, of descriptive differences of dialects, of historical stages of development and of basic differences of languages in different cultures. Finally, on a more theoretical level, the linguistic tradition was related to the studies of the neighbour disciplines logic, rhetoric and philosophy. Fundamental descriptive structures and conceptualizations were developed and expressed in systematic linguistic theories that were considered as the backbones of linguistics.

In the next chapter we shall carefully consider and reflect on the systematic principles that led to precise analyses of and methods for the descriptions and systematizations of established forms. The first chapter of this part concentrates on formalist perspectives in which *grammatical theory is based on description of formal structures and structure differences*. But some decades of focussing on structure analyses also provoked specific controversies. The critics required that the linguistic theories should not focus solely on formal mathematical or logical structures of syntax. Instead, studies of syntax should be integrated with studies of semantics, which should increasingly be given *priority*. Nevertheless, apart from these controversies between formalist or the usage-based linguistics, it is characteristic for both approaches that the central aim is language study, that is, language as it is used and considered as regular by adult speakers of each language. The first idea is that there are fundamental morphosyntactic or semantic structure principles that develop

[2] Givón (1995) presented an interesting start in the section "On the Co-evolution of Language, Mind and Brain" the last chapter of his book *Functionalism and Grammar*.

necessarily in language development of our ancestors or in language acquisition of children. The contrasting idea is that language evolution, development and acquisition generate necessarily language competence of intermediate stages of complexity. The elements of intermediate competence remain available even in adulthood, at least in non-standard communication when addressing children and also in speakers of basic language varieties who are developing speakers of a pidgin language. It is not yet confirmed that infants learn necessary foundations of competence in basic semantics and pragmatics during the first 18 months of their life. I believe that *language description of early stages of development* is important and provides a particularly fruitful research area in which linguistics and neurocognition can cooperate. In one of the following chapters this perspective of linguistic approach will be explained in detail.

In addition to presenting each of the different systematic orientations of form and meaning in their own "logic" the following chapters will also clarify their differences. Motivated by my intention to correlate linguist theory with models of neurocognition, I sympathize with each of the approaches, at least in some respects. As I shall explain in Chapter 8 the formalist approach has advantages in constructing translations of formalist syntactic models into functional neurocognitive interaction networks. On the other hand my theoretical intention is also to discuss possible models that encourage reflections about *how syntax and semantics become integrated* in the processes of development and acquisition. In this perspective I favour the perspective of layered developmental dynamic.

5.2 Chomsky's traditional base

In the second half of the past century modern linguistics was given a new impetus by Chomsky's (1957, 1963a, 1963b) generative grammar. After much enthusiasm about Chomsky's formalist approach many linguists developed more extended perspectives of analysis and understanding, increasingly following basically different orientations and analytic formats. Rather than representing the spectrum of differences this chapter concentrates on one particularly fruitful development, as Jackendoff (2002) presented it in his book *The Foundations of Language*. The next chapter will instead present a contrasting alternative. Instead of presenting formal linguistics the discussion will concentrate on some aspects of usage-based linguistics well presented by Langacker (2008) in his book *Cognitive Grammar*, supplementing it at some points by other analyses, as for instance Wierzbicka (1999), *Emotions across language and culture*.

The best entry into this chapter's topics is made by following Chomsky's own words from *Aspects of the Theory of Syntax* (1965): "The investigation of general grammar can profitably begin with a careful analysis of the kind of information presented in traditional grammars. Adopting this as a heuristic procedure, let us consider what a traditional grammar has to say about a simple English sentence." We must remark that the description strongly relies on specific grammatical categories and on general structure terms such as string, function, structure description, abstract structure representation, tree graph etc.

As an example Chomsky selected the sentence

(1) Sincerity may frighten the boy.

His first analysis was as follows:

> The string (1) is a sentence (S); *frighten the boy* is a Verb-Phrase (VP), consisting of the Verb (V) *frighten* and the Noun-Phrase (NP) *the boy; sincerity* is also an (NP); the (NP) *the boy* consists of the Determiner (Det) *the,* followed by a Noun (N) *boy;* the NP *sincerity* consists of just an N; *the* is, furthermore, an Article (Art); *may* is a Verbal Auxiliary (Aux) and, furthermore, a Modal (M).

This is the first list of grammatical descriptions. It concerns the sub-division of the string (1) into continuous sub-strings, each of which is given a certain syntactic category.

There is a second list in which

> *sincerity* is said to function as the Subject of the sentence,
> *frighten the boy* functions as the Predicate of this sentence,
> *the boy* functions as the Object of the Verb Phrase.

Here subject, predicate and object name, together with some other terms, the classical functional notions.

A third list enumerates, among others, the following statements

> *The boy* is a Count Noun (as distinct from the Mass Noun *butter* and the Abstract Noun *sincerity*),
> *sincerity* is a Common Noun (as distinct from the Proper Noun *John*, and the Pronoun *it*) and also an Abstract Noun,
> *boy* is furthermore an Animate Noun, and a Human noun,
> *frighten* is a Transitive Verb, in the Present Tense and Active Voice, taking the Progressive Aspect freely, allowing Abstract Subjects, and Human Objects.

These are so-called *sub-categorizations*, which indicate, together with their syntactic functions, some intuitively understood semantic features.

Together, all three lists exemplify how a typical traditional grammar may assign grammatical content to an English sentence. Chomsky acknowledges that the traditional information is, without question, substantially correct and is even essential to any account of how the language is used or acquired. Still, he is not satisfied. The new orientation of linguistic analysis should rather demonstrate *how information of grammar could be formally presented* in precise structural description. I shall postpone the discussion of these basic aspects to Chapter 8 continuing now to present the concrete perspectives exemplified by specific sentences and words.

Let me briefly refer also to Chomsky's earliest but simple sentence in his famous ground-breaking book *Syntactic Structures* (1957).

(2) The man hit the ball.

Neglecting for the moment the modal auxiliary *may* in sentence (1) the correct list of word and phrase categories as in (1) can be directly obtained by merely substituting *sincerity* by *John*, *frighten the boy* by *hit the ball*, *frighten* by *hit*, and *the boy* by *the ball*. The resulting sentence is

(3) John hit the ball

On this level the parsing structures of (1) and (3) are the same. The same holds for the second list of syntactic functions; the substitution generates the correct assignments of subject, predicate and object. The sub-category list is slightly different. Whereas the Subject *sincerity* is a Common Noun, the Subject *John* is a Proper Noun and an Animate Noun; the Object *ball* is a Common Noun and an Inanimate Noun, whereas the Object *boy* is a Common Noun and an Animate Noun. Finally, *frighten* is a Transitive Verb, allows Abstract Subjects, does not freely permit Object Deletion, but takes Progressive Aspect; *hit* is also a Transitive Verb, in Past Tense and Active Voice, that preferably takes Animate Subjects, permits Object Deletion, and takes Progressive Aspect.

These lists might be quite confusing for the non-linguist. In any case merely reading them does not help systematic transparency. Preferring formats of abstract analyses Chomsky invented a tree graph representation, a structure representation, which is shown in Figure 5.1. This tree graph demonstrates how appropriate graphic representations are important for scientific research.

Before turning to more basic feature descriptions here are again the perspectives implied in the lists. The first enumerated categories for Words (parts of speech) as well as for groupings or combinations of words, which are called Phrases. The features of the second list have an entirely different status. Whereas the part of speech of *sincerity* is Noun (N) the second list names its functional status, namely being the Subject of the sentence. The phrasal category of *the boy* is Noun-Phrase (NP), but its functional status is being the Object

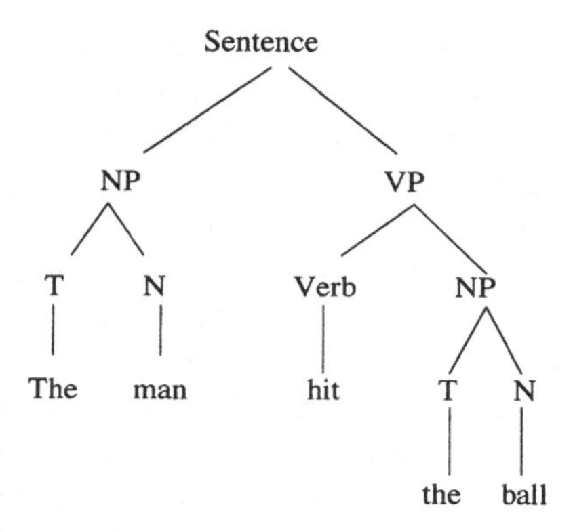

Figure 5.1. Simple syntactic structure.

of the Verb-Phrase. The Verb-Phrase *frighten the boy* has the functional status of Predicate of the sentence. The third list presents Sub-categorizations, using the terms Count Noun, Common Noun, Abstract Noun, Animate and Inanimate Nouns, Human Noun, Animal Noun, Transitive Verb and Forms such as Active or Passive Voice, Present, Past, or Future Tense, Progressive Aspect, Imperfect and Perfect Aspect. Let me note, that the present description belongs to a type that the previous chapters called descriptive and static. Dynamic aspects of language development that can be studied in children, in history or in evolution are neglected in the usual grammatical frameworks.

Let me briefly address the neuroscientist. He may accept that structure is presented in terms of a system of categories but would he ask whether we have to assume that this is the form in which grammatical information is represented in the cortex? Is it possible, that the representations of categories and sub-categories are somehow distributed and connected in neuronal units? In a more sceptical perspective, the non-linguist may suspect that the linguists' specifications of syntactic groupings, functions and sub-categories might just be their descriptive tools helping language teachers to teach their pupils to learn languages at school. Written grammatical terms may possibly address the activity of some areas in the pupil's brain, but the results of their learning will be quite different: a fine-tuned neural configuration organization whose structure is very different from what we see on paper. Acknowledging that second language learning by adults is partially guided by the traditional terminologies and categorizations on paper brain structures finally organize their more or less fluent competence as not a grammatical symbol manipulation but as a neurocognitive network.

But does this proposal by cognitive neuroscientists provide better ideas of how the brain organizes language competence? In section 2.3 of Chapter 2 I suggested that the areas of the cortical brain in which the knowledge of phonology, syntax and semantics are organized, might be conceived as consisting of *overlapping networks of linguistic cognits*, potentially representing pieces of "language competence knowledge". Each cognit is a distributed relational configuration of local neuronal assemblies or modules. In a concrete application of acquired competence knowledge in a momentary situation-focus – for instance a sentence utterance in a situation context – the distributed collection of neural units belonging to the cognit's configuration turns into a state of synchronized activity and keeps this state about a few hundred milliseconds. This could be the case in a moment of understanding a single word. If afterwards a sequence of other structural knowledge units becomes activated, one unit after the other, as for instance during hearing the words of a sentence,[3] a *collection of the cognits* for the different words and form categories generates a connected collection of local neural assemblies. In this case the resulting activity might keep, for a short time, the state of simultaneous activity that corresponds to the cognitive elements of the sentence, mostly without any consciousness for their presence. In some procedures the internal neuronal causality defended by Fuster (2003, p. 226) and Damasio (1999, p. 39) might generate temporal order of sequential structure units, representing the activity of the structurally differently marked word and grammatical category knowledge units as for instance involved in grammatical trees.[4] In this perspective, configurations of knowledge are not relations of symbols on paper but rather networks connecting local assembly units (in the case of word-forms and meanings) or networks of such networks, either simultaneously or in causally determined temporal order. In the following discussions I will often indicate when the terminology of formal symbol configuration could perhaps be reinterpreted in terms of cognits.

5.3 Chomsky's formalist syntax base and its critics

As already indicated earlier, words, phrases and sentences should be understood as meaningful units in which grammatical structures of sound patterns are related to structures of concrete meaning.[5] Some classical terms enumerated

[3] Recall Friederici's 2002 measurements of auditory perception of time intervals of auditory perception of different components of sentences discussed in Chapter 3 at the end of section 3.5. Friederici's measurement-based schema is also introductory, presented and discussed in Baars (2007b, p. 338).

[4] See Figure 2.1 in section 2.2 of Chapter 2. Neural network structures of this type will be presented and analysed in Chapter 8.

[5] Thus connecting cognits in phonological and syntactic processing areas of the brain to concrete meaning representing cognits organized in many other brain areas that do not belong to the typical areas of language form processing, i.e., Broca's and Wernicke's areas.

in Chomsky's list indicate a semantic origin. It was widely understood, however, that the correct understanding and presentation of the interdependencies of form and meaning, or of syntax and semantics, are very difficult. The master himself drew the conclusion that the theories of languages should first *concentrate on syntactic structure* only. He was persuaded that the foundation of languages' complexity is basically characterized by formal syntactic structure.[6] Other levels of linguistic complexity, such as phonology, semantics and pragmatics should be *derived from syntax*, both in their descriptive and in their explanatory role. Linguistic research should postpone detailed analyses until the structure of syntax has been satisfactorily determined for particular languages and for the notion of language in general.

The open-minded linguist was persuaded that, in spite of the obvious and impressive differences of languages, a systematic analysis of fundamental properties of language form could discover at least a *common structure core for all languages*. He thought that, similar to theoretical solutions in other areas of scientific research, it should be possible to define a basic framework in terms of mathematically formalized structures, in particular in the area of syntax. Core knowledge of this type would *transform linguistics into a science*, perhaps even corresponding to the effect that proposals of Kepler, Galilei and Descartes had for physics as a natural science.[7] But Chomsky was also aware of a fundamental difference. He understood that the range of *Cartesian ontology* of matter was based on *measurable properties and relations of spatio-temporally extended things*, whereas ideas determining *his own ontology of mental competences*, comprising thought and language, would have to be based

[6] He followed a tradition of R. Carnap (1937) *The Logical Syntax of Language*. London: Routledge and Kegan Paul, in which already Carnap eliminated left the phenomenological and logical structure of meaning presented in his book *The Logical Structure of the World*. Chicago Ill.: Open Court (in German 1928). The possibility of defining a relational system in terms of a dynamic network analysis, was proposed by Fuster, Damasio was still out of sight, or, in any case, considered to be less scientific.

[7] Kepler was persuaded that the structure of the planet system could be expressed by mathematical structures, hypothetical movements on geometrical structures, as he believed. Galilei introduces the idea of mathematical laws of nature. Instead of conceiving a unified event of movement, he separates the description into two components, one a spatial trajectory, the other a time-line, and represents each moment of the movement as a correlation of a point on the spatial trajectory with a point on the line of time. Measuring the distanced allowed the definition of the laws of movement. (For both, see Historisches Wörterbuch der Philosophie ed. J. Ritter et al. Basel: Schwabe. Vol. 3 p. 502). Descartes made the essential step. He invents the notion of today's well-known three-dimensional coordinate system with potentially infinite extension. Each point of the space can now be uniquely denoted by a triplet of numbers. The single things are extended parts located in the three-dimensional space. Since the extension of time is also expressed by a time coordinate it is possible to define physical kinematics as a science based on arithmetic. Physics and mechanics are now marked by a search for mathematical laws of nature.

on other types of mathematical knowledge. The challenge of finding a linguistically appropriate framework was overwhelming.

He understood that the principles of formal logic and of formalist mathematics strictly relying on formal symbol structures would perhaps provide an appropriate foundation. Already in 1937 Carnap had presented his logical syntax of formal languages. His student Bar-Hillel (1950 and 1954) proposed applications to natural languages, and Chomsky (1963) adapted the descriptive mechanisms of logic to certain established syntactic structure representations that had become familiar in the domain of American structuralism.

On this base various developments of theoretical linguistics called generative grammar were very successful and influential. But during the past decades many schools of linguistic studies no longer believed in "syntacto-centricity" of linguistic studies. Instead, there was, and still is, a growing conviction that syntax and pragmatico-semantics – and certainly also phonology – should be directly studied in their specific correlation. In order to simplify the conceptual intuitions most approaches dispensed with formal structure analysis, often with interesting results. On the other hand some of the new positions are quite open for discussion about how neuroscience, neuropsychology and psychological phenomenology could help to clarify concrete linguistic descriptions of language use.

In principle I share the views that semantics and pragmatics must be assigned a central role of analysis in addition to syntax. The following chapters will discuss the more extended connection of formal syntax with formal semantics and more intuitive functions of meaning. They will first concentrate on aspects of cognition in the sense I just discussed. The perspectives of the previous chapters, Fuster's working models for the perception–action cycle and Damasio's ideas will be particularly relevant, thus contributing to advanced forms of meaning-semantics and knowledge integration, in particular in the last sections of this chapter as well as in Chapter 6.

5.4 Jackendoff's three stages of organization

I will now return to approaches that consider gradual stages of modification correcting pure syntactocentrism to integrated analyses of comprehensive language organization. I will concentrate on the three stages presented in the very interesting and fruitful analyses of *Jackendoff's* book *The Foundations of Language*.

As his *basic idea* he introduces the notion of *f-mind* (functional mind in linguistics), as a first modification of Chomsky's main idea that abstract structures are *mental phenomena*. He insists that an appropriate analysis of mind should be determined by a framework in which the discoveries about brain properties should have a more direct bearing on *functional* properties, as was previously thought. Here the term *functional* implies a relation of mental structure to brain structure. Looking to the other side the f-mind linguist hopes to find partners among neurocognitivists who cherish the development of *functional* neuroscience. The functional neuroscientist is interested in explaining how certain characteristics of brain structures determine mental functions, and vice versa. Thus f-mind studies of linguistics should aim at finding neural structures that could be understood as counterparts of linguistic structures, frameworks and plausible principles. And neurocognitive structure analyses[8] should operate within a functional neuroscience explaining the brain's architectures and processes in terms of f-mental distinctions of phenomenological functions, for instance linguistic structure organizations, frameworks of interdependencies and principles of integration. I fully agree with this basic position. The function of mind is determined by reference to structures of brain constitution and understanding the function of brain components requires, already in the definitions and interpretations of test procedures, reference to types of mental understanding of acts and knowledge! Considering this interdependency, Jackendoff's notion of f-mind is welcome. It signals a move towards understanding mutual functionality whose formal foundation is elaborated in three stages of theoretical linguistic framework modification.

The *first stage* of modification claims that an f-mind approach must define general structure principles of a syntactic *core system* for language description: The core of each language is to be presented as a *combinatorial system*. Powerful combinatorial systems allow the definition of formalist sub-theories of language in the style of early generative grammars schematically illustrated above. Jackendoff explains the underlying ideas of this stage as follows: Language is a communication system in the natural world. The primary communication act is executed by sound utterances. One of the most striking facts is that speakers of a language can create or understand an unlimited number of different utterances about an unlimited number of topics. This productivity is possible thanks to an important design feature of language: utterances are built by combining elements of large but finite vocabulary into larger meaningful expressions. Language-specific expression forms are determined by combination principles based on a set of grammatical features and categories.[9]

[8] For instance Fuster's cognit structures which have been discussed in the first part of this book section 3 in Chapter 1.

[9] Formal details about combination principles and their translation in neuronal networks will be explained in Chapter 8.

Obviously, this phrasing appears fruitful to a formalist who represents combinations in terms of formal symbol configurations, for instance strings and formal tree configurations. In this stage Jackendoff does not yet explain how such structures could be related to empirically justified neuronal networks in the brain. In Chapter 8 I shall propose some initial answers to this question.

The *second stage* insists on substituting syntacto-centric architecture of expression combination. Instead of specifying a single combinatorial structure for syntax the new framework for language analysis contains an arrangement of *multiple parallel sources of combinatoriality*, one for phonology, one for syntax and one for semantics. In more detailed analysis, each domain may still be sub-divided into tiers. The three domains and sub-domains are mutually related by rich systems of interfaces.[10] Each domain represents phenomena existing in the world of objects: structured sound events, structured groupings of sound events and hierarchies of groupings of groupings, and finally conceptual representations of things or events that may be understood as single units, occurring combined in situations, or being parts of things or events. For each of the three domains, obviously hinting at phonology (in the wide sense, including prosody), morphosyntax and semantics, there is a domain-specific structure-determining system. This idea has two consequences: First, each structure-determining system is formally defined by a formation rule system. Even semantics is, on this stage, a formalist generative system, defining the abstract conceptual part only. Concrete perception–action semantics is still neglected. Second, since the formation rule systems are domain-specifically different, it is necessary that the parallel systems phonology, morphosyntax and meaning rule semantics be interrelated by special organization forms, so-called integration systems.

Jackendoff's *third stage* leads to a radical change of perspective. His earlier theorizing was based on the crucial idea that semantic theory is to explain how reference and truth-value are attached to linguistic expressions. Reflecting deeply however about an f-mental theory he came to a radical conclusion. He had already accepted that an f-mental account of language implies that language itself is organized in the brain. Thinking about semantics he is now certain that "it is necessary to thoroughly psychologise not just language but also the world"

[10] Jackendoff (2002, p. 207). He still accepts Carnap's structure principle discussed above. He writes that the outcome of his new orientation is a theory of grammar that is true to the fundamental goals of generative linguistics. Each parallel sub-discipline, phonology, syntax, and semantics is presented as a rich combinatorial system (p. 124). Formally he specified the parallel architecture as follows: "The overall architecture of grammar consists of a collection of generative components G_1, … G_n that create structures S_1, ... S_n, plus a set of interfaces I_{jk} that constrain the relation between structures of type S_j and structures of type S_k." This parallel architecture affords a clearer integration among the sub-fields of linguistics and between linguistics and related disciplines (p. 107).

(Jackendoff 2002, p. 294). His main arguments rely on situation-determined pointing phenomena that are relevant even when the observer has no conceptual clarity about the situation or when what is said does not at all match with the normal conceptual truth of the sentence expressed. The first case is exemplified by pointing to an object moving behind a shade, possibly a kind of an insect, and simultaneously addressing a communication partner by emphatically saying, "What is that?" There is no clear concept whatsoever, no reference nor truth, concerning the something that might be behind the shade. But still, the pointing act reaches its intention of arousing the partner's attention. The second example is as follows: Sitting on a bank, a girl says to her boyfriend "What a beautiful sunset." Both know, that in a strict sense the sun does not set but merely *appears* to be going down. Still, the girl's remark is completely correct and efficient in the situation, despite the fact that she knows, and knows that he knows, that the truth is rather that the earth is rotating. I think that Jackendoff's conclusion is that our competence of the appropriate use of language depends on given situations. Sometimes truth and conceptual object identifications are relevant, sometimes not. Consequently Jackendoff addresses the new intention of the third stage by the heading of his section 10.4 "Pushing the world into the mind", in view of the mind/brain analysis. Recalling our studies of the previous chapters, we should come to the idea that the selectivity of the pre-frontal cortex in non-standard situations could explain the organization of the concrete perception–action situations.[11] In any case the strictly logical analysis of thought is too restricted for ordinary language. Jackendoff's *third stage* must eliminate logical analysis when situations mark it as inappropriate. In principle we must look for working models of *functional mental networks realized* on various scales of detail by functional brain networks.

I agree with these ideas but think that Jackendoff's constitution of the mental networks system is still too limited. In my view, the problem of his conceptualist model, here represented in Figure 5.2, shows that, in terms of organization the third stage *still* understands the functional brain components as defined by *combinatorial systems*, now intended to represent a system of mind/brain organization.[12] I think that the principle of combinatorial organization is still misleading since it relies on manipulations of concept arrangements. We learned, however in the previous chapters, that cognits, as soon as they are active, form

[11] This is similar to Fuster's idea of cognit networks (Chapter 1, section 1.6).

[12] Jackendoff (2002, p. 304) circumscribes the conceptualist theory of reference by the following statement: "A speaker S of language L judges phrase P, uttered in a context C, to refer to entity E in [the world as conceptualized by S]" and contrasts it with the circumscription of the realist theory of reference: "Phrase P of language L, uttered in context C, refers to entity E in the world."

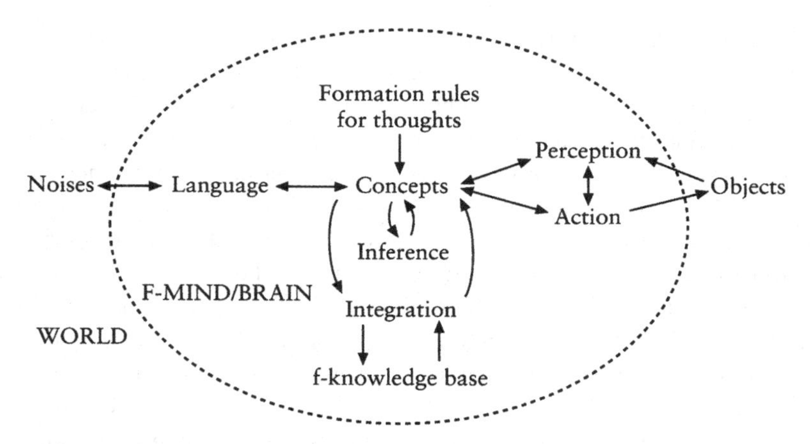

Figure 5.2. Pushing the world into the mind.

binding structures of interacting networks, one for each cognit that contributes to constituting the binding structure. Neurocognitive binding is not a combination of symbol patterns, as I will demonstrate in Chapter 8.

Still I fully agree with Jackendoff's intentions to find the way from mental analyses to descriptions of corresponding brain structure systems. The world must be understood in our brains. Jackendoff is correct in rejecting critiques such as "We don't perceive our percepts in our heads, we perceive objects out in the world" (Jackendoff 2002, p. 30). This statement is rather global. It is obvious that we need our head to perceive and that there is perceptual organization in the brain. The organization processes lead definitely to terminal states that are neural representations of percepts. But there is more. The *brain* brings us also to feel that what we see, i.e., the objects of our perception, is not in the brain but "out there" where we see it. But how do we know that it is located "out there"? As already explained above, neurocognitive processes of seeing generate an internal map that directs the eye, the pointing hand, and the grasping hand towards particular locations, when we want it. Still, it is possible that we err when in fact we only think that there is what we see. In fact we know that this situation may happen and even may be quite probable. Our brain therefore developed *checking procedures*, whose positive result contradicts the momentary assumption of a possible error. If the situation remains unclear the checking procedure may rely on judgements of other people to obtain increased security of perception.[13] In other words, the brain's organization of checking

[13] Here Jackendoff (2002, p. 331) refers to Searle's notion of *joint intention* (1995), a notion which Searle later, in (1998, p. 121), called *collective intentionality*.

perception procedures supports and controls brain maps locating perception–action directions of possible operations towards objects in the world. For me, the objects I see in the world are really there insofar as critical checking brain organizations, confirming that the intentional content generated by the interplay of visual brain processes, are satisfied.

To recap, I agree with basic ideas of Jackendoff's three stages, but think that, based on neurocognitive knowledge, a *fourth stage* is necessary, empirically as well as philosophically.[14] Jackendoff's third stage model is essentially based on sets of abstract "objective world model constructs" together with sets of abstract "formal symbol pattern constructs". My *suggestion of a fourth step* leads to a basic revision of complex meaning, properly taking account of the neurocognitive framework of mind/brain/body. Chapter 7 will be dedicated to explaining this improvement of "pushing the world into the mind/brain/body".

5.5 The world of thinking and knowing, loving or hating, happy or sad mind/brain/bodies

Though we do not yet know exactly the details of our mind/brain/bodies we know that all human beings are meant. This fact creates a problem for Jackendoff's project, here and, as we will see later, also with certain grammatical views that he accepts. We learned already from Damasio (1999) that mind/brain/bodies have a self, and learn to understand others, their thinking and knowing, loving or hating. The human capacity to attribute wishes, feelings, and beliefs to other people was studied by several neuroscientists who gave it the name *Theory of mind*. The phenomena are also known by the names of *mentalizing* or *empathizing* (Blakemore and Frith 2005, pp. 97–99). The consequence for Jackendoff's (1983) project of "pushing the world into the mind" should be clear. First, in the intended project the notion mind must be named in the plural: The world is intended to be pushed into mind/brain/bodies each of which is a mind/brain/body similar to the others in the sense that each can mentalize or empathize others. All other entities of the world are in this respect different from the own mind/brain/body. If this is so, there is another consequence that must be drawn; mind/brain/bodies must be basically distinguished from other entities. The ordinary physical entities that can be identified in the internalized world must be fundamentally differentiated in each mind/brain/body from the human individuals.

[14] It directly corresponds to the same kind of methodological solipsism which Carnap (1937) considered as his epistemological base.

There is another particular property that must be assigned to each mind/brain/body: Each of the basic perspectives refers to existence, involving continuous activity. Minds activate thinking and speaking, brains are activated by dynamic cells, and bodies have, as we learned from Cannon (1939), "their own wisdom" organized by the autonomic nervous system. All three are *living entities*. This is the point at which you may recall section 1 of Chapter 2, the presentation of Leibniz' and Spinoza's points of view. You may also recall that Damasio (2003) refers to Spinoza even in the title of his last book.

In other words, as soon as we follow Jackendoff's project certain *principles of abstract analyses must be revised*. Whereas logic, mathematics and other formal sciences tend to take entities as basic elements of formal analysis that only secondarily may be distinguished by abstract sub-classification, a "world pushed in the mind" *must distinguish in principle the other minds from the other animals*, but more essentially different from other objects that are neither mind/brain/bodies nor living beings. The semantics that necessarily follows from Jackendoff's project would be incompatible with a mere sub-classification of the class of abstract-defined entities. This may at best be a secondary solution when an analysis prefers to apply abstract descriptions instead of one based on the "pushed world".

Not only the modern semantics must be revised but also the syntax. Here the necessary revision is the so-called *argument structure* and the idea that a *transitive verb expresses a simple relation* with several arguments. The idea was adapted in linguistics from formal sentence analyses that were introduced in formal logic, in particular by Frege (1879). The next section will briefly explain Frege's proposals and the necessity of revision.

5.6 Frege's proposals for sentence analyses

Jackendoff is one of the linguists who followed the Fregean proposals. He writes that every theory since Frege acknowledges that appropriate representations of word meanings, in particular of verbs, may contain variables that are satisfied by arguments expressed elsewhere in the sentence. Jackendoff's first example of an argument structure formally representing the semantic core of a sentence, is *the lamb devoured the lion*. Its formal core is presented by the verb representation DEVOUR [x, y] where *the lamb* satisfies the argument x and *the lion* the argument y. He later summarizes again that every theory of semantics back to Frege acknowledges that word meaning may contain variables that are satisfied by arguments expressed elsewhere in the sentence.

The famous logician Frege originally introduced this structure into logical formula analysis. As a first step Frege proposed that a mathematical statement presented as a formula expresses in fact a thought that is either true or false. He next moved from mathematical formulas to language analysis in discussing the transitive sentence *Caesar conquered Gallia*. Here the sentence represents the content of a thought, in this case one that is true. Informally one would be tempted to say that the statement is true because Caesar did something in 50 BC, namely organizing based on his excellent military competence and his Roman army, a war in Gallia that was completely successful for him and ended with a complete victory.

Though meaning theory was not Frege's interest and his original analyses were concentrated on the logical foundation of systems of functions of arithmetic expressed in formulas he discovered that arithmetic equations are in fact statement expressions and thus have a language form. He claims that any language form contains a sense content and that each sense content of a statement expression has in general a truth value, meaning that it is either true or false. This principle holds for arithmetic equations and it holds also for linguistic statement sentences as many modern linguists claim.

In the next step Frege introduces the essential point: In general it is possible to conceive a statement-equation or statement-sentence separated into two parts, one of which is complete as an expression, whereas the other requires that an expression be added. As an arithmetic example Frege presents the equation $(x+1)^2 = 2(x+1)$ and explains that the equation statement is incomplete. To transform it into a complete statement the variable x must be substituted in both places by the same rational number.

Replacing x by the numeral 1 leads to a complete statement whose truth value is **true**. Replacements by any other number leads to complete statements with the truth value **false**.

Correspondingly the statement *Caesar conquered Gallia* can be separated into the two parts the first of which is the incomplete statement marked by a variable:

x conquered Gallia

Replacing x by a noun leads to a complete statement. The statement resulting from replacing x by Caesar is true, whereas the replacement by General Grant leads to a false statement.

Caesar conquered Gallia Grant conquered Gallia

Frege adds a remark that is very important from my point of view. The generalization of the procedure from *equations with variables* to *statement sentences with variables* assumes that variables do not only take *numerals* for completion but also *objects in general*. At the same time he acknowledges that this implies the problematic idea that "a person is counted among the objects". But for developing the application of his proposal he insists that any objects without restriction should be allowed as values in functions, even in generalized types of functions. If for instance in the incomplete statement

The capital of *x*

x is replaced by Germany it becomes clear that the completed expression is not a statement expression that is either true or false but rather a naming phrase whose value is Berlin.

Here again, Frege acknowledges the problem of the question: "What at all can be called object?". He answers with a radical generalization: "Object can be called everything that is not a function", i.e., the content of a formula expression with one or several variables, but rather what is referred to by an expression without a variable. Since a complete statement sentence does not contain a variable, it denotes an object! As a consequence truth-values true and false are objects! It is obvious that this definition of an object is completely formal. As such it contradicts the common sense notion of object as it is explained in dictionaries. But formalist linguists like Jackendoff use Frege's formal notion of object. As a consequence in any complete expression, for instance names, like thunderstorm and angels replacing *x* or *y*, in the semantic formula DEVOUR [*x*,*y*] denote objects. This also holds for Jackendoff's other example sentence

Beethoven likes Schubert

For the purposes of semantics Jackendoff even uses a new descriptive unit, namely *typed variable* that distinguishes types of variables whose details will be discussed in Chapter 7. A complete semantic formula for this sentence results from the following general variable constituted schema

Z [*x*,*y*]

by replacing the general variables by typed variables, *Z* by EVENT, *x* by OBJECT_A and *y* by OBJECT_B, that in turn are replaced by LIKE, BEETHOVEN, and SCHUBERT. Thus Beethoven and Schubert are semantically typed as objects, obviously still in Frege's formalist sense. In contrast to Frege Jackendoff does not explicitly acknowledge the problem of conceptually typing a person as an object.

He probably also accepts another conclusion of Frege's proposals that became fundamental in formal logic. It implied a revision of a basic element of traditional syntax. According to it the fundamental split of a sentence is into the components of subject and predicate. The predicate may be the combination of the verb with objects (in the grammatical sense) or of the verb with a predicative. An example for the latter is *Jack drinks wine, Jane water*. In their meaningful status *wine* and *water* are *predicative* and *not objects* of the sentences.

Let me summarize: A strict understanding of Jackendoff's project of pushing the "world into the mind" has, in my view, fundamental consequences, in semantics as well as in syntax. Let me repeat: Whereas logic, mathematics and other formal sciences tend to take entities as basic elements of formal analysis that only secondarily may be distinguished by abstract sub-classification, a "world pushed in the mind" *must distinguish in principle the other minds from the other animals* – partially mind/brain/body similar – but more essentially from other objects that are neither mind/brain/bodies nor living beings. Before studying them in Chapters 7 and 8, it might be fruitful to study a variety of linguistics that is more open-minded for principled criteria that characterize the use of language.

6

Grammar as life

6.1 Explaining grammar as meaningful

In the previous chapter I presented linguistic studies that concentrated on the integration of syntax and conceptual semantics. Chomsky (1957) proposed several varieties of formal structure descriptions originally based on standard frameworks of traditional grammar. Jackendoff (1983) demonstrated that an appropriate integration of phonological, syntactic and semantic phenomena requires a reorganization, in which each of the different domains determines its own principles of formal structure description. We have seen that this insight led to *three stages* in which the domain's structure descriptions were represented independently but systematically integrated by means of interface relations. The *previous chapter finished with a radical revision:* The foundations of different domains and their integration should no longer be represented as phenomena in the body external world, that is, the world of things. Instead the external world should be pushed into the mind/brain/body that organizes the ranges of internal feelings and externally oriented perception, action and objective thought.

The reader may ask why many modern schools of linguistics had widely accepted logical formats as guidelines for descriptive representations. He should recall that antiquity and the middle ages already understood grammar and logic as related disciplines. All other disciplines had to be understood in domain-appropriate conception frameworks following schematic knowledge of grammar and logic. Accounting for the enormous progress of mathematics and logic linguistics also aimed at correspondingly improved formats and theories that nevertheless should be adapted to the characteristics of *natural* languages. Without abandoning the formalist techniques, theoretical linguistics should clearly circumscribe and define the basic properties that distinguish ordinary language from other notational systems, say of computer science, the genetic code, the "language" of bees etc. or from other knowledge frames defined for the sciences physics, genetics, formal information theory, computer theory etc.

As mentioned in the previous chapter an increasing number of linguists were however generally dissatisfied with the formalist frameworks. Careful analyses made it clear that a language should no longer be limited to defining the systematic of structured expressions and their formally structured interdependency. The new orientation, called usage-based linguistics, insists that grammar must demonstrate how the speakers and hearers make use of the enormous efficiency and flexibility of possible expressions. Langacker (2008), an influential representative of this movement, used the impressive slogan: "Grammar as life". In his view, the speakers of a language select spontaneously phrasings and structures from the repertoires of possible expression and "construe" what they consider as appropriate in the communicative situation. And in almost all communicative situations the hearers also reconstruct a situation-appropriate understanding of language form and meaning, considering and evaluating the implications they have, momentarily or in principle. The mentally acquired grammar of speakers and hearers offers them an enormous number of variations, differently adapted to the background of external situations and interpersonal understanding. Serving the functions of concrete usage a grammar description is integrated into communicative life. This is how Langacker explains usage-based linguistics, summarizing that "portraying grammar as a purely formal system is not just wrong but wrong-headed".[1]

Langacker's main perspective is to explain *grammar as meaningful*. Instead of restricting lin guistic analysis to syntax as a precisely circumscribed *core* of language, as Chomsky emphasized, his studies rely primarily on semantics together with studies of grammatical forms supporting communicative efficiency. I agree with his intention though not with some of his basic external object-oriented framework archetypes. Langacker acknowledges that such studies must be grounded in activities of the brain, which functions as the integrative part of the body. But unfortunately, his studies do not further explain the brain's involvement in appropriate analysis of language usage. Thus it remains our task to reflect about the brain's part in explaining certain aspects of Langacker's cognitive grammar. I am satisfied that, in principle, Langacker implicitly acknowledges that the *brain activity* is based on a complex biological system that should ultimately instantiate psychological functions, especially in phenomenological understanding of semantics. But here again I regret

[1] Langacker, R.W. (2008), *Cognitive Linguistics*. Oxford: Oxford University Press. But in certain studies of translating grammatical organization into hierarchies of functional neurocognition formal structure will be very helpful, as some sections of Chapter 8 will show.

that his present linguistic attitude still feels forced to analyse "grammar of life" relative to the external *world we live in and talk about*, disregarding their foundation in our mind/brain/body organization. I will now discuss some basic aspects involved in Langacker's proposals.

6.2 Langacker's view of the foundations of grammar

As already said "Grammar as Life", the first headline of Langacker's book (Langacker, 2008), marks the directions of his research. Let me briefly sketch what he means. In contrast to the tradition he does not believe that grammar must remain based on abstract principles unrelated to more extended aspects of cognition or human endeavour. But he goes too far when criticizing formal grammar systems as wrong-headed. Studying precision of fundamental core phenomena of grammar is useful in view of a comprehensive understanding of grammar. Chapter 8 will concentrate on these aspects and will also show how in our present situation of research formal systems can be related to details of neurocognitive explanation, for instance Fuster's (2003) cognit system. Langacker's claim would only be justified if the formalist approaches neglected the fact that their working models apply only to core knowledge, thus not acknowledging that their descriptions would definitely need to be complemented in more comprehensive description formats and domains, such as semantics and pragmatics, as well as by interface descriptions which relate form and meaning.

In this extended perspective Langacker's main goal to explain grammar as meaningfully integrated in a framework of informal semantics is very fruitful. This orientation can and should indeed be pursued. In my perspective the framework would have to be based on mental and neurocognitive conceptualization, explaining in particular the notion of communicative efficiency.

In this respect, Langacker's principled position of linguistic meaning is problematic. But on p. 4 he acknowledges that a mental phenomenon, conceptualization, is grounded in physical reality, the activity of the brain, which functions as an integral part of the body, which functions as an integral part of the world. Langacker's statements clearly characterize the primary perspective of his philosophy: Mind depends, via brain and body, on the objective world. But in later sections this view is complemented by perspectives in which linguistic meanings are also grounded in social interaction, being negotiated by interlocutors based on mutual assessment of their knowledge, thought and intentions.

Obviously, this is also a reference not only to an objective world, but now the world of social facts. The position seems to be similar to Saussure's (1878) foundation of linguistic structure, Broca's discovery of language in the brain, but systematically only established by social interaction of individuals. This position rejects studies of individual minds and brains because it is not directly concerned with the objective foundation in social conditions.

There is a fundamental error in Saussure's and in Langacker's view. Studying the individuals' brains, in particular their environment-based development during infancy and childhood, necessarily demonstrates the constitution of social knowledge in the individual's brain. Probably together with Jackendoff, I would complain that Langacker's philosophy of a fundamentally objective view neglects the development and the role of the individual's acquisition of self-knowledge integrated with the other-self knowledge implying a social foundation. Those of us who concentrate on this view when "pushing" the material and the social world into the mind/brain/body do not neglect access to external facts and to a social understanding. We would consider both as derivatives of the collective mind/brain/body organization.

In my view, considering the *diverse aspects of neurocognition* would lead Langacker, as well as the other usage-based linguists, to further aspects of linguistic relevance, namely the body-based vivid mutual evaluations, emotions and feelings of people, their knowledge of self as well as the complexities of externally oriented perception and action systems. Reference to neurocognition is necessary. The perspective differs fundamentally from one merely computationally determined by formal interaction with formal reasoning about external things, situations or mere considerations. The perspectives also differ from neurocognition in concentrating on merely usage-based analyses of the world. Langacker's argument is that we live in the world and talk about it. There are more dimensions of vivid and evaluated experience, many of them extending far beyond the objective reality that is claimed to be their external foundation. In its analytic and descriptive power our life expands in developing imaginative phenomena and mental constructions and in internal feeling. Nevertheless, I will show in this chapter that the critical discussion of Langacker's proposals is fruitful and suggests extensions of Jackendoff's world in the mind/brain. But in his neglect of neurocognition Langacker is also wrong-headed.

6.3 Mental and communicative efficiency

Language expressions serve two basic functions: communication and internal thought. In formal logic or mathematics rigorous argument and proof may completely rely on explicit statements of relations between abstract formulas or abstract expressions only. Not so in natural languages! Here, the general principle of usage is *efficiency in meaningful acts* of communication or practical and narrative thought. Ordinary language develops grammatical forms of speech acts in serving this principle. The ultimate competence of adult speakers of a language consists in large sets of common sense knowledge they share with the hearers. Many types of knowledge, in particular those that are relevant in situations and in conversations about concrete situations of life support the efficiency of communication.

The expressions and conversations become integrated in a spontaneously active mind/brain system. They combine organizations of grammatical form, meaning knowledge and evaluation of perception and efficiency of action under control of attention and intention. Their combination integrates momentarily efficient expression-forms and meanings. Often, there are also underlying "mechanisms" that speakers' and hearers' brains are able to put to proper use in given moments, without being able to say how the brain networks function. Following our discussion of Fuster's models in Chapter 2 we may assume that complexes of mentally triggered competences and neuronal networks correspond on various levels of detail.

Our understanding of the brain's system is still limited, though not quite as reduced as our common sense understanding of the operation of a TV. Here we learn to manage the different functions by pushing the keys but we do not know how the activated electronic networks satisfy our requirements. But as children it was not even necessary to mechanically learn some key usage. As children our neural networks connected perceptions and actions of the "key words" during the periods of our language acquisition.

In some sense our knowledge of how to relate grammatical form and meaning was and still is automatic. We are tempted to say that they represent the mental knowledge and usage conditions of meaningful talk and thought. There is, however an important practical feature: Very often words and normal sentences do *not convey "technically" detailed meaning*. They merely activate a more or less global intuition. Let us take the meaning of the word *walk*. Everybody can be assumed to have a practical knowledge of walking. Hearing someone using this word the hearer's mind/brain activates the practical knowledge similar to the speaker's. But *what really* does the word "walk" mean? Looking up some dictionaries we shall read such explanations as "move along fairly slowly by putting one foot in front of the other on the ground" or " move along on foot in a natural way, in such a way that one foot is always touching the ground". Both explanations are *not really precise*. The first would not exclude skating or sliding and the second does not exclude that these would not be natural ways of moving along. The speakers and hearers of language know that spontaneously. But there are other aspects of reduction. Even they do not provide more precise physiological details nor details of mind/brain activities such as those organizing the body's equilibrium, the correct sequencing of the legs etc. Still a user of the dictionary is normally satisfied. He would not be more satisfied with a complete explanation of all physiological details that would take many pages and perhaps even book length.

Wegener (1885), who was a schoolteacher during the nineteenth century, exemplified the phenomena of linguistic efficiency with reference to Diomedes walking to combat and argued that detailed description is certainly inappropriate in poetry, and even in ordinary language. Thus Homer did not tell us the details of Diomedes' walk. How did the hero really move his legs while walking to combat, how did he pay attention to the stones on the way? Wegener continues to ask whether language is required to provide fully detailed description. He argues that according to the principles of language's usage, detailed description might only be considered where the speaker may assume that the hearer does not at all know what *walking* means. But even then it would be more informative for the hearer if the speaker *demonstrates* him an act of walking. The bards of ancient times may have accompanied their speech by singing and mimicking by dance. Wegener insists on the efficiency principle: *We should not go beyond certain limits of descriptive detail*, limits that take the presumable knowledge of the hearer into account as well as limits that distinguish the domain of non-conscious purely "automatic" activity from conscious goal-directed movement. For example, it may be assumed that sequences of movements caused by muscle activation such as movements of the legs, eyes, head, arms must be considered as *ultimate components* of common usage of natural language, components for which no further descriptive specification is appropriate. A particularly typical case is *seeing*. The details of the organization of the saccadic eye movements during visual perception have clearly proven the enormous gap that exists between meaning concepts and our understanding on the one hand and their underlying organizational foundation. Completely different but of the same order of complexity are the notions of feeling and of the self, own self experience, other-self experience and each-other self experience, as discussed in Chapter 4, sections 4–6. It seems to be clear that this principled efficiency constraint of descriptive meaning can be applied to practically all verb meanings in natural language usage.

Wegener even proposes a plausible explanation: The meanings of common verbs may involve many components. But their combinations have been built and mechanized into automatic complexions available in many years of mostly non-conscious usage. Consequently the combinations are only directly accessible as a whole and not in their details. This is at least the case in *common everyday language*. A student of robotics may be interested in the details of organizing stable two-leg walks. He would study *physiology and some psychology*. But the scientist's explanations would not be easy reading nor very relevant in the perspectives of everyday usage. Today neurocognition aims at even more details whose elementary feature combinations enormously transcend psychological and physiological structure knowledge. Whoever is interested must accept more difficult reading. The *common sense principle of obeying the communicatively relevant limits of details* is an important semantic condition for everyday language.[2]

The principle is almost trivial for internal body-related experiences and feelings. As Damasio wrote (2003, p. 87–88):

> The immediate substrates of feelings are the mappings of myriad aspects of body states in the sensory regions designed to receive signals from the body. Someone might object that we do not seem to register consciously the perception of all those body-part states. Thank goodness we do not register them all, indeed. We do experience some of them quite specifically and not always pleasantly–a disturbed heart rhythm, a painful contraction of the gut, and so forth. But for most other components, I hypothesize that we experience them in 'composite' form. Certain patterns of internal milieu chemistry, for example, register as background feelings of energy, fatigue, or malaise. We also experience the set of behavioural changes that become appetites and cravings. Obviously we do not 'experience' the blood level of glucose dropping below its lower admissible threshold, but we rapidly experience the consequences of that drop: certain behaviours are engaged (e.g., appetite for food); the muscles do not obey our commands; we feel tired.

It is obvious that the explicitly worded characteristics are much more global in the case of feeling than in describing external experiences.

The existence of this efficiency principle and its automatic application in utterances is not generally taken into account. Jackendoff (2002, p. 123) claims for instance that semantics is the organization of those thoughts that language

[2] It also seems to be related to *aspects of memorizing*. The meaning of what somebody said in discourse is usually not memorized by all the details of the speech act (as in learning by heart) but by keeping "key" features of an event. It may be that it is in view of memorizing that the field of perception (Langacker's field of awareness) is reduced into conscious data that become conceptualized in memorizable arrangements. In the moment of perception the other data are neglected and thus become momentarily sub-conscious. (However, in this case the data may become conscious in careful attention and observation.) The *momentary sub-consciousness* must be distinguished from *reduced memorized consciousness* the stable form of sub-consciousness. This kind of sub-consciousness must in turn be distinguished from facts that become strictly inaccessible, such as for instance the regularity of grammar (including the formal understanding of the sameness of categories (noun, verb, prepositions and their arrangement). Grammars become a part of advanced knowledge analysis, just as other specialities of disciplines and science (Searle 1992, pp. 83–93, 95) that are not part of an ordinary language speaker's background.

can express. This is not true. Common language accessible for everyday usage *does not* include all thoughts that language in the more general sense *can* express. Linguistics is concerned with common language for everyday usage. Consequently *linguistic semantics must be concerned with that part of situation understanding and background knowledge comprising the dominant feature of everybody's everyday experience. This is the foundation of efficient conventional communication*: *Semantics is concerned with that part of thought that refers to established things fitting the limits of everybody's everyday experience framed for efficient conventional communication.* Efficient selection may select among pre-structured alternatives.

Much richer domains of thought are accessible in *extended formats of common language*. We may think of complement enriched formats of everyday language applied, for instance, in advanced literature or in terminologies and regimented formats of scholarly disciplines. Here we may even find special symbolic forms of representation such as formulas, figures, schemata etc. These extensions show that language can express many more things than efficient communicative limits of everyday language formats would allow. In general, advanced disciplines usually require complete expression of relevant details.[3] Common sense communication requires efficient expression forms that would not express what could be assumed as obvious to the hearer.[4] The common sense knowledge of word meaning usually takes this into account. In this way ordinary language is particularly *efficient* in everyday communication of adults.

The *principles of efficiency* are already learned very early in proto-grammar. Givón gives a list of some principles of efficiency constraints. The following are quantity rules:

a. If a word meaning is not expressed and the situation predictable it will be left unexpressed.
b. If a word meaning is not expressed but relevant it will be left unexpressed if it is unimportant in the given situation.

[3] Even here, this is not quite correct. A physicist carefully eliminates properties that do not seem to be relevant for the laws he tries to discover. If he studies mechanical dynamics he will assume that the colour of the particles can be neglected in his studies.

[4] This is a point at which another discipline, namely philosophy, deviates from ordinary language communication. Its principles are particularly well explained by J. Searle, an expert of transparent presentation of philosophical argumentation. See J. Searle (1998, pp. 157–161). Transparent argumentation in philosophy does not deviate by special terminology but rather by asking questions and demanding deeper explanations of words and concepts that normally appear to be sufficiently understood in ordinary language. These facts are non-conscious in the sense of the ordinary speaker.

There are moreover phonetic and syntactic rules characterizing pre-grammar or proto-grammar that are not directly relevant for our discussion of meaning (Givón 1995, pp. 406–407). But in historical development of language there is a change of criteria of what is normally considered to be unimportant or irrelevant. In the historical development of narratives and explanatory discourse another direction of speech organization develops. The addition of morphemes to lexical units becomes influential. Adding morphemes, clitics and particles helps to guide the understanding of the hearer. These phenomena of grammaticalization allow developmental flexibility of language without interfering with the efficiency provided by the basic word order and word combination. Givón writes that the rules of proto-grammar are found intact in more grammaticalized language. Nothing is lost, but rather a considerable amount of grammatical machinery is being introduced, advanced abstract rules, grammatical morphology and complex hierarchic syntactic constructions generating more flexibility in communicatively evolved grammar.

But combinations of efficiency with flexibility transmitting communication content do not only occur in ordinary language usage. New language formats develop by serving special purposes of cognition or specific socio-cultural conditions. *Extended formats of common language* are generated. The analysis of the latter does not belong to linguistics but should still be considered in the present context. Formal logic and the studies of formal foundations of mathematics should rather be understood in contrast to ordinary language determined by efficiency of everyday communication. Since extended format languages serve special frames of intelligent thought they follow their own criteria of efficiency. These criteria aim at supporting precision of thought.

But it is also interesting that scientific language usage can lead to retroactive influence of terminologies and schematic representations on specific conventions of spoken expression usage. It is well-known that passive constructions and other impersonal clauses used in scientific texts are rather marked as uncommon in oral communication. Unmarked common language rather uses active clauses with common words for expressing events and situations. Thus there is more variation of style depending on communicative, socio-cultural, cognitive, literate or scientific frameworks of usage. We come to understand that the criterion of communicative efficiency of common usage changes with the complexity of involved frames of knowledge and elements of form. Learning the use of narratives and explanatory discourse requires the adding of morphemes, clitics, and particles that help to guide the understanding of the hearer and to provide more flexibility in communicatively evolved grammar.

6.4 Flexibility of grammatical framing in constructions and construals of form and meaning

As just explained, stages of complex regularities of language structures unfold advanced systems of knowledge expression leading to efficiency of language formats in ordinary language usage as well as in advanced disciplines of language-based intelligence. Unfolded precisions of form indicate the possibilities of languages' flexibility. But the following idea is misleading. It is not true that unfolding language form, namely the complexity of grammatical structure, and of syntax in particular, is the necessary condition for expressing efficiency, flexibility and precision of knowledge on each level of thought. Instead there are two fundamentally different *stages*. During the *first years of life* early components of language structuring are framed by neural organization of perception–action systems of memories and networks organizing early forms of intelligence. After some years, complexity of language structure begins to provide structural support as well as fluency of narrativity, and argumentative thought and accessibility of advanced archetypes of background knowledge emerge. Advanced fluency and knowledge accessibility gives language structure increasing priority in processes of advanced thought.

Concentrating linguistic studies on the fundamentally advanced stage of language led to an idea cherished in particular in formal sciences such as logic and mathematics: In language priority should be given to explicit expression of precisions in selecting formal expressions and formal meaning relations between terms. The truth of this idea is limited. In ordinary language efficiency of expression has priority.

In ordinary language phenomena of *grammaticalization* allow developmental *flexibility* of language without interfering with the *efficiency* provided by the basic word order and word combination. Givón (1995) writes that the rules already acquired in primitive proto-grammar are not eliminated when language provides usage of more complex structure. Instead the primitive knowledge is found intact in more grammaticalized language. Nothing is lost, but rather a considerable amount of grammatical machinery is being introduced, advanced abstract rules, grammatical morphology and complex hierarchic syntactic constructions generating more flexibility in communicatively evolved grammar.

As just explained, language usage does not only serve the purpose of momentary efficiency *and economy* in the usage of utterances. The other criterion is *flexibility*. Early proto-language in children – as probably also in proto-historical language – tends to be economic in the usage of words. Proto-sentences rather rely on *skeletal expressions*. Together with the development of narrativity or primitive forms of argument, more secure understanding of what was said requires however *more detailed construals*. The prize of the primitive level is that processes of partial *rigidification* and *ritualization* control conventional expressions.[5] In more developed language complexity a primitive skeletal base expression *girl likes boy* may receive a *grammatical grounding and variation* as in Langacker's examples *the girl likes that boy; this girl may like some boy; some girl liked this boy; each girl likes a boy; a girl will like the boy; every girl should like some boy; no girl liked any boy;* and so on (Langacker 2008, p. 259). It should be obvious that these phrases receive their momentarily specific function in contexts of situations or narratives. The functions of the grounding elements are as follows: Through nominal grounding (e.g., *the, this, that, some, a, each, every, no, any*), the *speaker directs the hearer's attention* to the intended discourse referent. Clausal grounding (e.g. *–s, -ed, may, will, should*) situates the profiled relationship with respect to the *speaker's current conception of time and reality*.

From the *developmental point of view* bounding does not only use morphemes bound to nouns and verbs. There are also particles, such as adverbs, prepositions and conjunctions that play in general similar semantic roles as modifiers. Their syntactic roles may well be different (Jespersen 1963, pp. 87–90). It was also suggested that distinct cycles of symbolization complexity took place in the evolution of human language. The first involved the evolution of a well-coded lexicon for nouns and nominals, verbs and verbal idioms; the second the evolution of grammar based on sequential word order and hierarchically organized syntactic construction. As already mentioned the proper integration of bound morphemes, clitics and particles, together with an efficient distribution of pauses and intonation, lead ultimately to fluent utterances of simple and complex sentences generated in brain organization.

In the *traditional descriptions of grammar* one concentrates on the constitutions of word-forms, clauses and sentences composed from their parts and one investigates the constraints as well as the regular possibilities of composition or combination. What does not violate the constraints and what is possible according to rules can be called a *grammatical construction in* the language. Various forms and representations of constructions have been developed in the tradition. In the first perspective and in common sense understanding one thinks that the constraints and rules are rigorous; such as whether a given word-form or clause sentence is correctly composed or not. It is known that speakers often seem to deviate, but to common sense this only means that nobody is perfect, not even in his competence of language. Some people even think that

[5] Particularly interesting observations about ontogenetic and early historical development are found in T. Givón (1995, pp. 10, 360n, 402, 440, 441).

in a correct language there is exactly one expression that properly expresses each given fact of the world. In logical calculi, in mathematics and in rigorous sciences this is true in principle.

But in ordinary language the appropriateness of language usage must not be strict grammatical correctness. Sometimes the efficiency of an expression depends more on the *context of meaning and rule* or what the speaker intends to say, what is the given situation given the environmental and social sense etc. Thus it is obvious that the *dominant requirement* of language is not to be rigorously correct but to be *flexible* and communicatively *efficient in context and situation*. With the help of the pre-frontal cortex a speaker's brain does not select what is formally correct but what the given situation requires to be *construed as an efficient utterance*. It is true that irregularity is not frequent. But the reason is that the brain organization blocks irregularity for the reason of inefficiency.

An expression that is most appropriate to communicative intention in a communicative situation is the one that most efficiently renders what is meant. The selection of that expression is the *act of construing*, which is accounting for properly selected relations of meaning and expression forms, and constitutes efficient and clear *construal*. Every adult has learned that there are more facets to the structure of a complex expression than can be represented in a single constituent structure.

The flexibility of language seems to be very complicated and it is, as every learner of a language realizes. But already trivial examples can show the phenomenon, for instance Langacker's introductory example for a *construal*: *There may be water in a glass*, an archetypal content that may serve as the underlying meaning of the following construals, which provide different perspectives: (1) *the glass with water in it* designates the container; (2) *the water in the glass* designates the liquid it contains; (3) *the glass is half-full* designates the relationship wherein the volume occupied by the liquid is just half of its potential volume; (4) *the glass is half-empty* designates the relationship wherein the volume occupied by the void is just half of the potential volume.

Construals are indeed core phenomena of Langacker's usage-based approach. It is obvious that an appropriate study of Langacker's system would confront us with too many details. Let me merely summarize some essential aspects of his approach. Contrary to the generative grammarians' concentration on levels of basic core structures and their formal interfaces, the analyses of construals try to show how variations of expressions provide flexibility of linguistic form and meaning. The essential point is how flexible construals deviate from systems of grammatically standard expressions, the so-called constructions. In this view the system of constructions provides a language's sub-system of schematized expressions. It comprises the default representations and interpretations. We learn from Langacker that, by emphasizing *usage-based linguistic frameworks,* his cognitive grammar concentrates on the notion of construals that mark the possibilities of linguistic variations in which the same situation can be expressed. The expression variations are considered to be a particularly important factor called *profiling*.

In Langacker's view the important factor allowing variability is *semantics in the broad sense, including pragmatics*. In this perspective Langacker's principled methods depend primarily on aspects of meaning on two levels: a meaning consists of both *conceptual content* and a *particular way of construal*, that is the characteristic view in which content can be presented.

6.5 Objectivity and subjectivity in common forms of situation accounts

The application of the efficiency principle leads to selections of aspects that are properly expressed in the construals of sentences and texts. This fact is often related to a certain feature of vision. Consider a visual situation in which the attention focuses on certain features or concepts in the range of things and events that are visible *but are not selected by attention* to be *consciously conceptualized* as a *perceived fact*. In Langacker's cognitive grammar the notion conception subsumes perception. At the same time visual perception is used as a paradigmatic metaphor for the efficient representation of situation ranges in terms of linguistic expressions. A meaningful linguistic expression is an enhanced form of conscious perception, whereas conception-oriented selection may or may not become conscious.

Rather than exposing a general schema of all details I shall follow a typical example of Langacker's book. Here also he prefers a focused perception of visual situation.

> Imagine yourself in the audience of a theatre watching a gripping play. All your attention is directed at the stage and is focused more specifically on the actor presently speaking. Being totally absorbed in the play, you have hardly any awareness of yourself or your own immediate circumstances. This viewing arrangement therefore maximizes the asymmetry between the viewer and what is viewed, also called the subject and the object of perception (Langacker 2008, p. 77).

Next Langacker also transfers the notions of *subjective* or *objective* in a situation in which the "viewer" of the mentioned event reports to a friend what he saw yesterday. Now *two persons are talking to each other*, the previous "viewer" now being the speaker. In Langacker's terminology the viewer is the *subject of conceptions* to be properly *profiled in* expressions. In the situation of communication he is the conceptualizer (the primary *communicative subject*) in his narrative and his attention selects what is to be conceptualized (that is the primary *communicative object*) in a communicatively appropriate profiling. (The selected expression's profile must serve the task of generating a sequence of expressions that appears to be appropriate to guide the attention of the second subject, namely the hearer. He should be prompted to generate second objects, i.e., the hearer's imaginative understanding of the objects of attention. Steps of profiling the story's meaning are selected by the speaker's constructive and the hearer's re-constructive attentions. The profiled meanings are brought to mental existence in maximal intentional "*objectivity*" (attention focus), expressed in Langacker's terminology. He emphasizes that, on the other hand, both, speaker and hearer, are construed in maximal "*subjectivity*" since they function *merely as subjects of conceptions*, that is by their "tacit conceptualized presence whose real counterpart is not itself [linguistically] conceived". Langacker declares speaker and hearer to be part of the *conceptual substrate* supporting an expression's meaning. In the range of their conceptualized meaning, their *personality is not in focus*. The substrate is also called the *ground* of the speech event.[6] In principle the substrate or ground is usually not explicitly expressed by words. Nevertheless, the ground must not always remain verbally tacit. It may be profiled: Words like *I, you, here* and *now* may be used more typically for facets of the ground. But considering the range of possible foci, the meanings of these words remain off stage and are construed with only *minimal objectivity*.

Langacker systematizes (2008, p. 261) the underlying process with reference to the functioning of the eyes:

[6] The term substratum introduces a very important notion in the philosophical perspective. It is the Latin translation of Aristotle's term *Hypokeimenon*. In many translations it names the logical subject, understood as underlying the specifics of the predicate. Related to this interpretation Hypokeimenon is also understood as the foundation of becoming and passing away. It also is related to the notion *substance*. I would not assume that Langacker wants to imply one of these meanings. Very probably he rather wants to take the term in accordance with a dictionary definition, for instance "substance, which lies beneath and supports another." I will later return to the discussion of these aspects. But I must confess that I have something in mind, that is well-founded by the meanings of Hypokeimenon in the sense either of a logical subject understood as underlying the specifics of the predicate or as the foundation of becoming and passing away, and a term related to substance in the classical sense. All three aspects are closely related to the foundational notions in Leibniz' philosophy. In my book from 1991 *Leibniz*' perspectives of ontology and language were indeed foundational but at the moment it is sufficient to apply the basic ideas to the models of Jackendoff and Langacker in which they indicate the "full scope of awareness" that in some sense could also be understood as maximal scope.

> In vision the perceiving subject is the viewer … as well as the *subjective locus of experience* inside the head (the *mentally constructed perspective point* from which we 'look out' at our surroundings). At a given moment the full scope of awareness consists of everything that falls in the visual field, and the onstage region is the portion presently being attended to. The object of attention then is the focus of our visual *attention* – that is, the onstage entity specifically being looked at. The eyes are construed with maximal subjectivity, for they see but cannot themselves be seen. What they see when examined up close and with full acuity, is construed with maximal *objectivity*. Construed with a lesser degree of objectivity is everything else currently visible, both onstage and offstage. The scope of awareness even includes parts of the viewer's own body, which is vaguely perceptible at the very margins of the visual field (Langacker 2008, p. 77).

This analysis is particularly interesting due to the large number of mind/brain functions that Langacker now uses. We find the notions of attention, awareness, subjectivity and objectivity. The terms could perhaps be used as bridges to the neurocognitive descriptions of pre-frontal attention control (Fuster 2003) that selects activity patterns within the perception–action systems, whereas profiling might indicate inhibition of other neurocognitive features that are not in the focus of the selective organization. In fact, however, the words merely present global indications of what is seen and, perhaps, how it is evaluated. Other features are lacking. Neither the "viewer's" emotions nor features organizing specifics of memorizing and of memory access are explicitly mentioned, though they certainly contribute to additional specificity in language organization.

In spite of the general and interesting openness of the framework I have reasons for some *critique*. I will prepare it by referring to the *two Langacker stories*. The first is about the visitor of the theatre, the second about him in the situation of reporting the performance *to his friend, the hearer*. This communicative example was intended to show in this particularly central example that speaker and hearer are *shadowed elements* that merely serve the specific function of relating the presence-profiled person – the speaker in the communicative situation – to the inexplicit ground. What is conceptualized in the report is the visitor, present and even absorbed by the play, but not consciously present in person. Langacker calls this position *offstage* since the person is merely the onlooker. In a clear sense he is the subject of the perception act, but he is not assigned a specifically objective role. Thus Langacker characterizes the visitor as merely subjectively present, and not objectively, in the sense of objective presence of the actors on the scene. He claims that the visitor's *experience of his own subjective presence* is *kind of shadowed*, since he was fully absorbed by the events on stage. The same holds for the speaker in the second story. Here the visitor will tell the events to his friend without using explicit first person statements. Now comes Langacker's essential statement: In the communication both are only *"subjectively" present* as well as in the theatre where the speaker was merely the onlooker. Here is Langacker's general definition: Someone is *construed with maximal subjectivity in a situation or a sentence* in which he *functions exclusively as subject* who *sees or reports*. Langacker thinks that the conscious onlooker *lacks conscious self-awareness*.

Here is the critique: The account seems to be problematic in the case reported by the story. The visitor was declared as "absorbed" by the play. In this case one cannot assume that he would think, in recalling the situation, that he did not *experience being self-aware or self-present*. Considering Langacker's predicate" having been absorbed" in feeling, at least a component of his mind/brain/body – and hence of his self – was indeed fully engaged in what was going on.

Even in the second story, in which the speaker refers to the event's report to his friend the visitor may remark explicitly that *he felt very happy* seeing this play. He might even say that he was enthusiastic *to see how the actors acted in this or that situation*.

It may be that Langacker wanted to say that during the moments of perception the visitor does *not follow* his *own prior intentions, intended thoughts or intended acts* nor conscious control of situation observation. Taking, however, the perspective of the brain, the organization of the feeling of being absorbed is indeed very active. In other words, Langacker may be right as far as he follows the phenomenological analysis of externally directed perception and action, but he is wrong when neglecting the emotion and feeling organizing system based on the autonomic nervous system, as explained above in Chapter 4.

What I want to say is, that *Langacker's notion of subjectivity* is considered only in cases in which the cognitive state of a person is fully involved in executing perceptions, actions or speech. His externally oriented phenomenology leads him to consider only acts that are *externally directed embodied acts*, suppressing consciousness of the self's own state of feeling or the self's explicit experience of the body-based self-presence. Only exceptionally Langacker mentions in his book *intellectual or emotive* states of experience, or different types of mental constitution of *social, cultural, intellectual and emotive* dimensions.

6.6 Subjective and objective time

After some years of studies of generative linguistics I felt challenged by Montague's proposal to integrate theoretical linguistics in the strictly mathematical framework of Montague grammars. The formalism relied strongly on the classical notion of time. Thinking about the relevance of this position I studied various alternatives, for instance, the notions and thought of poets such as Valéry and philosophers such as Bergson, Husserl, and Wittgenstein, and was led to the question of how linguistics could be psychologized in a neural framework (Montague and Schnelle 1972; Schnelle 1981a, 1981b, 1981c).

Of particular interest was Husserl's clear contrast between subjective and objective time (Schnelle 1981a). He is quite explicit that objective time is a complex structure relative to which objects are conceived as positioned, a structure that is itself constituted by some complicated mental process on the basis of individual experiences. The precise nature of this constitution is not yet completely clarified. The clarification is the main challenge for the phenomenology of temporal consciousness.

The influential book of Miller and Johnson-Laird (1976) presented the scientific counterpart for clarifying the notions of time and tense. But I was not satisfied with their proposals. They still relied on the formalisms of logical analyses (Miller and Johnson-Laird 1976, especially p. 418), though they acknowledged that they do not regard temporal logic as a theory of the psychological processes that people must execute as they speak and understand English sentences but as an abstraction from such a psychological theory. I fully agree with their words: "When we understand the psycho-neural engine that keeps these temporal relations straight, we will probably find that it bears little resemblance to any logic machine." But I do not agree with their consequence: "However it realizes its functions will have to deal with the kind of information that we have here represented abstractly in terms of R.T and quantificational formulas. We are merely proposing rather abstract boundary conditions on the kind of psycho-neural engine that we can expect to find" (Miller and Johnson-Laird 1976, p. 458).

I must say that I rather preferred to study the development of more concrete models and descriptions of the 'psycho-neural engine', as it was for instance presented in Popper and Eccles' book *The Self and its Brain* (1977). A dynamic and interactive framework should also be applied to the notion of time and the analysis of aspect and tense.

From the linguistic side I found support from Grünbaum (1969) when he insisted, "that nowness and temporal becoming are not entitled to a place within physical theory". It became definitely clear that Montague's grammar and formal logical approaches are unable to represent "nowness" nor do they give a meaning to the "flow of time" nor to "becoming". It was clear that formal analyses alone would be inappropriate. Many detailed discussions with my wife, a photographer, historian of art and philosophy and analyst of visual experience, on perceiving paintings, pictures and the environment deepened my insight into experiences of perception. In particular I learned from her very much about the role of saccadic eye movements and the experiences of perceptual movement, such as for instance the generation of afterimages, assuming that the linear understanding of time, which we learn from our clocks and calendar, should have any relevance for our internal experience organization based on neural processes in the neural system.

It was obvious that only an interdisciplinary analysis that considers the interdependency of the internal experiences of psychology, biology and a new perspective of linguistic semantics would lead to a clear distinction of constructed and schematized knowledge of the world from the characteristics of our personal experiences.

Other linguistic help came from Chafe (1973 and 1974). He argued for a new analysis of tense that would provide new perspectives for 'psychosemantics'. In considering the role of temporal adverbs in sentences with the past tense, he was led to the view that sentences without temporal adverbs usually have a special status. He invites considerations of sentences like

(1) *Steve fell in the swimming pool*

He distinguishes three cases of use. In our context the third case, *the generic perfect* is important. Here the sentence is uttered 'out of the blue' with no time reference previously established.

Chafe claims that an utterance of a sentence according to case 3 "communicates, among other things, the information that the event in question is still fresh in the speaker's mind. Typically it is something perceived by the speaker recently enough so that it has not yet left his consciousness". He introduces the notion of surface memory relating to information kept in the memory similar to a short-term memory and emphasizes that "retention of something in the surface memory is not a function of 'real' or 'physical' time, but rather of what might be called 'Experimental time' ".

The reference to surface memory is not only communicated by past but also by *non-generic perfective*, such as

(2) *Steve has fallen in the swimming pool.*

It is usual, that in an utterance of a sentence like this, the speaker is not only concerned with Steve's falling in the pool, but also with some other event or state, obvious to the speaker or hearer, such that this other event is a consequence of Steve's falling in the pool. The non-generic perfective is relatively complicated since it refers to two events (instead of one) and a relation between the two. "If sentence (2) carried the generic meaning, it would convey the idea that Steve has fallen in the pool on one or more occasions during his life, the times being left unspecified."

These remarks are interesting but led me to a critical remark: Chafe's discussion is related to the ordinary use of tense in (an adult) having the competence of fully developed language. However it seems to me that the earlier use is instead undifferentiated. It is a pre-tense, the earliest time expression of a child. Some linguists seem to hold that the most primitive use of tense is strictly related to *aspect* or, rather, that *some forms of aspect are more primitive than tense*. Based on this insight, I conceived of developments of potentially dynamic networks that might explain the phenomena (Schnelle 1980b). In the summary of this article I wrote: " 'Now' and the impression of passing is an ingredient that seems to be fundamental in operative organisms since their actual state and their change of state not merely mark a now and the passing now but rather an experiential experience of the flowing of now."

I summarized that analyses of aspect and time should distinguish the following experiential differences: internal, external, constructive and technically controlled measurements:

1. immediate self-experiencing feeling of life (aspect of punctual duration of inner experience),
2. episodic memories located in contexts of other events,
3. subjective time "measure" (time passes slowly or quickly),
4. objective time measurements (based on measuring apparatus),
5. objective time order of events (defined in a logical or mathematical framework of classical physics).

6.7 The mental universe as a collection of archetype frames

The competence on the five stages of experiential difference are acquired in the development of childhood and are then established as common sense usage and finally educated frames for world knowledge. Many theorists and scientists believe that only the scientific accounts are true and are reliable when asked for truth. This may be the case in the frameworks of pure reason. But when practical reason is required, there are not rigorous frameworks but instead quite a variety of prototypes that are used for communication, thought and judgement.

Many prototypes may in turn be collected as practically useful schemata in classes of archetypes.

On certain cultural levels the educated people acquire some of the principles for archetypes. This is for instance the case for the measurable notions of objective time and of objective space and some related phenomena. They obtain dominance over the experiences of the inner self and body. This even leads to analyses of mental studies, as for instance the linguistic studies of tense and aspects. Even linguists are easily satisfied with scientifically clear abstract boundary conditions for their descriptions of concrete semantics. As a consequence their accounts are often only correct for language use in adult speakers.

Langacker was quite correct in referring to aspects of efficiency and to the relevance of archetypes helping to frame the understanding of sentences and discourse. Unfortunately he first insists on physicalistic object and movement-dependent archetypes, such as the billiard ball model. I would think that certain archetype distinctions learned already in childhood must be kept as basic. It is rather obvious that not the general notions of *things* and *entities* are foundational but the distinct archetypes of *People, Animals and Things. The latter do not have an internal biological self-organization*. Do we need more fundamental archetypes?

Let me briefly return to what our Western culture claims as objective knowledge with an established base in the sciences. In a sense this is true. In our cultural perspective we register that there are fundamental reflections leading our philosophy to the unifying concept of entity. Descartes' framework still has some relevance insofar as material entities are characterized by their *spatial extension, moving in the world and pushing* each other, in all cases in measurable ways that frame the conception of reality. Langacker's billiard model (2008, p.103) is an offspring of Cartesian ideas. This is also demonstrated by his statement:

> We think of our world as being populated by discrete physical objects. These objects are capable of moving about through space and making contact with one another. Motion is driven by energy, which some objects draw from internal resources and others receive from the exterior. When motion results in forceful physical contact, energy is transmitted from the mover to the impacted object, which may thereby be set in motion to participate in further interactions.
>
> and he summarizes that
>
> this cognitive model represents a fundamental way in which we view the world.

But obviously this physical model contrasts radically with our biological models, which are instead offsprings of ideas that both Spinoza and Leibniz developed in view of substituting the Cartesian model. This was as already emphasized in Chapter 2, section 1. Shouldn't we rather look for archetype frames of the later type for explaining experience in order to find a more appropriate fundamental way in viewing the world? It should be clear that understanding *the integration of life and its functional constituents should primarily consider one or several archetypes relating to the intellectually most advanced biological model, the human person*? Descriptions in terms of physical mechanisms and their usually schematic laws are definitely far from an appropriate account of human mind/brain/body and from understanding the notions of person and self. Correct linguistic semantics and pragmatics accept these archetypes. The challenge is to explain also their neurocognitive foundation in the brain's organization. As I explained in the first chapters, exteroceptive and interoceptive perception–action organizations are as important as their integration in the organization of emotions, feelings and of the own self and the others' selves.

Analysing typical features of emotive expressives should mark the most characteristic difference of the "neurocognitive perspective" and "billiard perspective". Langacker studies some of them in a later section. But as a typical linguist he just lists a number of words and phrases and some contexts of use. He summarizes that an expressive may

> profile at least in a narrow sense of the term. In general an expression's profile is the onstage focus of attention, objectively construed by definition. But at least from the standpoint of the speaker, expressives are not about viewing and describing onstage content. In using one the speaker is either performing a social action or vocally manifesting an experience – rather than describing a scenario, he enacts a role in it. For the speaker, then, the action or experience is subjectively construed. While an expressive evokes and calls attention to it, the prominence it thus receives is not that of a focused object of description. If we stick to the narrow definition, therefore, expressives are principled exceptions to the generalization that every expression has a profile (Langacker 2008, section 13.2.4. p. 475 ff).

Langacker seems to suggest that the profiling structure and semantics necessarily refer to the world and structure and mostly visual onstage perception as it can be described in external and dominantly objective usage terms. There is no consideration of how perception and action are organized in mind and brain. He seems to think that a linguist must be satisfied with this kind of description.

This goes as far as it does. Using expressives is, according to Langacker, a principled exception. Langacker's reason is that their semantic and communicative roles cannot be properly distinguished by "onstage description".

Langacker does not suggest any appropriate solution and is certainly not interested in the role that the f-mind and the brain might play. His archetypes belong to the range of objective events. Archetypes in the range of emotions, feelings and self-experience and other self-experience would transcend his collection of externally observable and objectively analysable facts. In this limited framework important aspects of internal human experience are excluded though there are many utterance phenomena in languages that need more extended analyses. My next chapter will indicate some ranges that seem to require interoceptive analysis rather than the usual exteroceptive behaviour-based studies.

But before turning to these new "archetypes" I would like to summarize the fruitful perspectives of Langacker's approaches. He correctly emphasizes the interdependency of grammar and life. He indicates languages' communicative efficiency and flexibility in the way they conceptualize situations and facts. In connection with his notion of grounding explicit expressions Langacker introduces the notion of grammatical skeletons. Obviously they play the role of *integrating core themata*. A neurocognitivist like Fuster would call them *symbols*. Their presence in the speech act guides profiling of efficient, flexible and sufficiently clear syntactic construals that concentrate on relevant meanings.

7

Integrating language organization in mind and brain: the world of thinking and knowing, liking or hating other mind/brain/bodies

7.1 The integrated mind/brain/body: a new version of pushing "the world" into the mind/brain/body of a person

The previous chapter finished by emphasizing the importance of archetypes, mentioning also that Langacker's dominant archetypes belong to the range of *objective events*. He acknowledged that archetypes in the range of emotions, feelings and self-experience and of other self-experience would transcend his collection of externally observable and objectively analysable facts. In this limited framework important aspects of internal human experience are excluded.

Jackendoff presents a number of arguments for a fundamental extension in which language, thought, perceived things and events in the world are organized in our mind/brain/bodies (Jackendoff 2002, p. 272–273, 305–306). The available *structures of the world do not exist independently of our mind/brain/body's organization* generated in mutual cooperation and communication among social groups of people. This is even true for science; theories and measurement techniques are invented, developed, applied and checked in scientific communities. Though there is continuous search for progress, scientific knowledge also is never completed. I agree with Jackendoff that in view of extending our perspective we must go deeper into psychology and neuropsychology of neural assemblies for storing and processing conceptual structures in terms of neural assemblies. They interact with other organization systems, thus generating the interplay and integration of perception, action, attention, selectivity, emotions, feelings and self-awareness and mentalizing the psychology and functional neuropsychology of others. The complete system is in continuous mental and communicative contact with the normal community.

There cannot be any doubt that the thought in the individuals' and communities' mind/brain/bodies is not a mere combination of unambiguous, situation-

and context-independent word meanings in sentences and discourse. On the contrary, each process of understanding selects meanings that fit to context, background and discourse continuity. The first sections of this chapter will explain several details. Considering single words, and in particular nouns, we must acknowledge that various meaning aspects must often be selected and composed from archetypes' specificities. In these selections classifications in the frame of *fundamental types* must be taken into account. In our organismic and neuropsychological aspects of dynamics the following *distinctions of fundamental types for nouns* are distinguished by degree of biologically self-based dynamic of the named units. In speakers, they are established in developmental perspectives of human knowledge and understanding[1]:

a) *inanimate*, i.e., mentally reduced self-based dynamic,
b) *animate*, i.e., mentally advanced self-based dynamic,
c) *person perspectives*:
 c1) self-conscious self-based dynamic,
 c2) other person mentalized self-based conscious dynamic,
 c3) naming ambiguously.

Insofar as nouns are used in the sentences their fundamental types should be taken into account. I am certain that the nouns used in sentences and discourse must be archetypically evaluated in the sentence construction, and I recommend following Jespersen's ranking of nouns in their roles of subjects, objects and other units in the sentence.[2] Similar to the notion of sentence topic, to be discussed in the next chapter, the fundamental meaning rank is important for the principled evaluation of the sentence. We learn which elements are without self-dynamic, have only reduced *mental* self-dynamic or refer to self-presence or mentalized dynamic of others.

Accepting Jespersen's ranking as a principle of sentence organization, giving obviously the first rank the dominant position of a noun in sentence expression, I would reject the logical idea in which the verb represents the core of the sentence and all other lexeme types, nouns, adjectives, adverbs and prepositions are values of variables dependent on the verb that is of argument structure. For many theoretical linguists, including Jackendoff, my position does not seem to be acceptable. Trying to defend my position I will

[1] The fundamental character has its origin in early childhood and keeps being understood during adulthood. Only certain fixations on the physical world views tended to eliminate our natural mind/body/brain-based distinctions.

[2] Jespersen (1963) p. 96–97 ff. Remark that on p. 97 Jespersen also demonstrates that nominalization may also lead to shifts of ranking, and thus to different forms of profiling in the sense of Langacker (2008), p. 259f.

discuss the details in criticising Davidson's logical structure analysis of sentences in the next section, since this view is widely accepted in semantic and formal linguistics. Before exposing my quasi-Jespersonian defence, let me briefly present my core elements in critical discussions of some of Jackendoff's examples.

7.2 Criticizing verb-centred meaning structures

Many theoretical linguists were persuaded by Davidson and accepted his arguments and the ensuing meaning structure schema. Disagreeing with the majority I think that Davidson's arguments are not conclusive.

The essential part of Davidson's argument is relatively short and may be presented in explicit form (p. 38 in Davidson 2006;the original is from 1967):

> Strange things going on! Jones did it slowly, deliberately, in the bathroom, with a knife, at midnight. What he did was butter a piece of toast. We are too familiar with the language of action to notice a first anomaly: the "it" of "Jones did it slowly, deliberately..." seems to refer to some entity, presumably an action, that is then characterized in a number of ways. Asked for the logical form of this sentence, we might volunteer something like, "There is an action x such that Jones did x slowly and Jones did x deliberately and Jones did x in the bathroom,"... and so on. But then, we need an appropriate singular term to substitute for "x". In fact we know Jones buttered a piece of toast.

Until this moment everything is correct. We might perhaps add that instead of the last sentence we might also say "In fact we know that Jones did something, namely buttering a piece of toast". We know about his action of buttering a piece of toast. Elsewhere in his article Davidson says that he does not like the gerunds. I think that whatever may be the reason the gerund is just what Davidson wants. The gerund construction is indeed a single term, and we might well consider it as the appropriate single term. Substituting "action x" by "action of buttering a piece of toast" and for "did x" by "did: buttering a piece of toast". We then would obtain: There is an "action of buttering a piece of toast" such that Jones did: buttering a piece of toast slowly, and Jones did: buttering a piece of toast deliberately and Jones did: buttering a piece of toast in the bathroom...

The singular term looked for is the gerund with predicative *"buttering a piece of toast"*.

But Davidson continues in another direction. Here is the continuation of his text: "Allowing a little slack [sic!], we can substitute for 'x' and get 'Jones buttered a piece of toast slowly and Jones buttered a piece of toast deliberately and Jones buttered a piece of toast in the bathroom .... and so on." The trouble is that we have nothing here we would ordinarily recognize as a singular term. Another sign, that we have not caught the logical form of the sentence is that, in this last version, there is no implication that any one action was slow, deliberate and in the bathroom, though this is clearly part of what is meant by the original. In our example the first requirement is satisfied when adding in each substitution also the variable x.

Davidson summarizes again the fundamental reason of the exercise "I would like to account for the logical or grammatical role of the parts or words in such sentences and with what is known of the role of those same parts of words in other (non-action) sentences. I take this enterprise to be the same as showing how the meanings of action sentences depend on their structure". Davidson attacks quite a number of problems. But our problem is the role of Jones. Is subject, namely Jones, the one who did it and does everything else depend on his doing it?

Consider the mind/brain/body we must assign it for all adjuncts to the mind/brain/body of Jones. It is the logic of the internal organization of this mind/brain/body that determines everything else and whatever the mind/brain/body can do on the basis of his knowledge. Once again in Leibniz' and Jespersen's terms, with which I agree, it is not the logical formalism but the biological mechanism of the brain that must be considered.

Davidson meets the challenge of some of his critics that his apparent success in many points is due to the fact that he has simply omitted what is peculiar to action sentences. He writes that he does not think so: "The concept of agency contains two elements, and when we separate them clearly, I think we shall see that the present analysis has not left anything out. First, the agent acts or does something instead of being acted upon, or having something happen to him". As a particular point he correctly adds that "perhaps it is a necessary condition of attributing agency that one argument-place in the verb is filled with a reference to the agent as a person. It will not do to refer to his body or his members, or to anyone else. But beyond that it is hard to go. I sleep, I snore, I push buttons, I recite verses, and I catch cold. .. No grammatical test ... will separate out the cases here where we want to speak of agency".

We determine agency derived from the usage in which language says that someone did it. Though intentionality is often a central organization feature it is not at all necessary from the linguistic neurocognitive or pan-organic philosophical stance. Sleeping, snoring and so on are verbs with which we can answer the question of what somebody is doing. In the linguistic sense this broad range is enough. But also our analysis of our organization of mind/brain/body and our insight in contributions of parts of the brain allow us to say that the mind/brain/body individual does something. Logicians may not be satisfied but neurocognitive analysis may well correspond to linguistic usage. I am sure that Davidson's previous reference to a person is central and marks also Jespersen's first rank in the hierarchy of ranks in the sentence. I could moreover refer again to the pan-organic perspective of Leibniz' philosophy that was presented and discussed in Chapter 2.

I think that we are not forced by Davidson's and other logicians' arguments to fix sentence constitution on the equilibrium of noun-phrases instead of finding the appropriate ranking of the noun-phrases and the adjuncts. Person reference has priority often also when the noun-phrase has the status of the direct object. Chomsky's old example "Sincerity may frighten the boy" is a case in point. I will explain structure variations in the next chapter in section 8.1.c. Some free word order structures that are demonstrated as normal in German illustrate the usage. In German, the object of Chomsky's sentence can be regularly expressed as initial topic:

(7) Den Jungen (acc) mag Ehrlichkeit(nom) erschrecken
(8) The boy (acc) may sincerity(nom) frighten.

My summary is that for each sentence the first rank must be found. If there is a subject noun-phrase that designates one or several persons they have the first rank, the ranking sentence constitution should normally be given this form.

In our discussion of Davidson's logical form we learned again that formalizations of semantics that have been designed for mathematical or formal logical systems promise clear and transparent forms of representations for these disciplines. Our normal understanding of the world and of ourselves is, however much more complex. In his chapter about reference and truth Jackendoff explained some aspects of complexity and argues correctly and very clearly that further research of linguistics cannot be satisfied with describing only the external world.

7.3 The stars in the world of geometry, Beethoven and Schubert in the world of music

The present section is not about Beethoven and Schubert as stars but rather about stars in the sky and a composer's relation to another composer. Both topics are introduced by central example sentences used by Jackendoff. These sentences are:

(1) The little star's beside a big star.
(2) Beethoven likes Schubert.

Presented as a mere example sentence (1) confronts us with the typical problem of isolated example sentences. We see that it is grammatically well formed. But what does it exactly mean? Implicitly the definite article at the beginning indicates that the sentence presupposes a preceding context that must have introduced some idea of the star as a topic of discourse. Still, it is not quite clear whether a real star in the sky is meant or, perhaps a little star in a picture book being shown to a child. The word *beside* may indicate that both the little star and the big star are in some sense perceived as being near to each other or being in direct contact. Thus the two together should be given the first rank. We may even present another sentence in which a pair must be given the second rank. The sentence is

(3) This small stone has room between two other stones

The following two schemata would represent the meaning structure. Here is the interpretation for sentence (2):

[LITTLE STAR, BIG STAR]: ARE BESIDE [EACH OTHER]

The two stars mark the topic and the relation expression determines, together with the object, the predicate. In a similar way we may interpret (3):

[SMALL STONE]: IS BETWEEN [TWO STONES]

In his explanation Jackendoff presents two descriptive logical forms for the semantic levels. As just presented the two stars or stones are presented in formally paired structures. In other words, both stand in a relation denoted by *beside*, respectively by *between* in sentence 3. On the *second form of semantic level* a picture shows two star figures in a short distance. This second level is named "*Spatial structure*". In later chapters Jackendoff develops and adapts some earlier ideas about descriptions of lexical units in frameworks of a conceptual system or a spatial system etc. In Jackendoff's presentations they rely on adaptations of logical structure analyses.

They are however not without problems. Let me concentrate on his spatial *structure proposal.* He emphasizes that the spatial structure determines the objects of the world with respect to their constitution in space. The system is geometric in character. Judgements and inferences having to do with exact shapes, locations and forces can be formulated in terms of the spatial structure system. These formalist aspects will be considered in the next chapter. Here I will briefly consider the geometric character in semantic interpretation.

The proposed description clearly contrasts with perceptual organization of spatial phenomena from the point of view of visual scanning discussed in connection with saccadic eye movements. Many perceptual processes are multimodal in organization, thus relying on information from other modalities, touch, proprioception etc. Instead Jackendoff understands his model rather as "encoding image-schemas: abstract structures from which a variety of percepts can be compared". The spatial system defines rather structures and arrangements of complete object shapes and locations in a geometric space.

This *abstract and perception-independent system description* may be useful and particularly appropriate for its combination with the conceptual property determinations of objects in the conceptual system. From a deductive point of view of world knowledge the idea seems to be justified. But Jackendoff wants to *construct the mind/brain foundation of world knowledge*. It may be true that school-educated adults have indeed developed mind/brain knowledge involving deductive judgement and inference. But I rather doubt that this is true for the majority of people. Even very intelligent people and even creative people do not immediately access their formalist deductive mechanisms.

A typical case is *Einstein*. As already described earlier in section 4.1 his creative thoughts did not rely on ordinary language and formula but rather on visual imagination and imagined feelings in his muscles. His reflection was about the *pre-scientific* thought of the foundation of the notions of space, time and material objects and the ways in which science has modified them and rendered them more precise.[3] The first significant accomplishment was the development of Euclidean geometry whose axiomatic formulation must not be allowed to blind us to its empirical origin (the possibilities of laying out or juxtaposing solid bodies).

Reflecting archaic perceptions of objects and object relations is in many respects still practically relevant for the common sense. Einstein concluded that the notion of space gradually developed in stages of the following understandings of notions: bodily object > spatial relation of bodily objects > gap between objects > space. It was Descartes who introduced space as a con-

[3] See Albert Einstein (1961). *Relativity – The Special and the General Theory*. New York NY: Three Rivers Press. p. 162 ff. My explanations are supplemented by parts in the more explicit German text in A. Einstein (1986) *Mein Weltbild*. Berlin: Ullstein pp. 140–142.

tinuum, and Newton felt forced to use the notion of space as an absolute entity for reference of his mathematical laws. This notion of space remained as the fundamental framework for physics, until it was revised by Einstein's theories. The continuum of interactive energy in the world substitutes the abstract space.

I think that Einstein is correct. Educated peoples' Euclidean geometry is not the proper model of space and should certainly not substitute our experiential studies of perception–action and experience-related understanding, in particular not when we accept Jackendoff's proposal to "push the world into the mind/brain". The inductive forms of Einstein's imaginative creativity of mind provide a better guide.

My next problem with Jackendoff's proposal is called forth by the *second example sentence "Beethoven likes Schubert"*. It is in particular Jackendoff's formal description of this sentence. Here also Jackendoff is influenced by his basic idea of meaning structure. The standard is still the formalism of argument structure in which the verb represents the core of relational structure, in our case Noun-phrase +Verb + Noun-phrase. Jackendoff sees no problem with denoting the meaning of the complete sentence as an event! Was it an event that Beethoven liked Schubert? In any case Figure 7.1 presents Jackendoff's meaning structure of the sentence.

The reflection of this sentence will help us again in further precision of Jackendoffs "pushing project". In the present representation its denotations are rather strange. The notion event was already mentioned. And now Beethoven and Schubert are both characterized as *objects*. Classifying the famous composers Beethoven and Schubert as objects does not seem to suggest that here the meaning of the verb is not based on a personal emotion, the more typical use of *to like* but rather expresses *Beethoven's intellectual esteem for Schubert.*

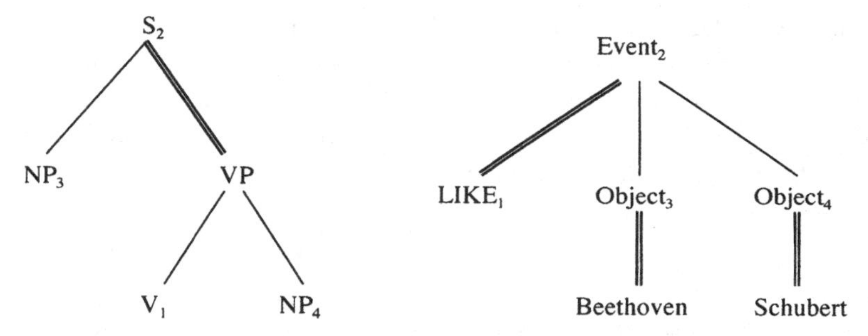

Figure 7.1. Syntax and semantic structure for "Beethoven liked Schubert".

Some details of the next chapter will have to explain this assumption and some details of the meanings involved.

Jackendoff introduced his last chapter, the one that contains our example, with the statement that it "is concerned with how the meaning of a phrase or sentence is composed, using parts of the meanings of its words". But in my account this statement will be given a sense that differs very much from Jackendoff's understanding.[4] Let me begin with the question whether the truth of the example sentence can be confirmed. Beethoven's biography indeed reports that several months before his death he said "Wahrlich, in dem Schubert wohnt ein göttlicher Funke." ("Really, in that Schubert lives a divine spark!") What may have been his reason? Since both are incredibly gifted composers, it is almost certain that Beethoven thereby expressed his great esteem for Schubert's compositions and not his personal feeling. In fact he never met Schubert personally. Thus Beethoven's liking Schubert must be interpreted in terms of a *specific archetype of creativity*, namely the one of musical composition. Interpretations of more common meanings of the word *like* would here be either empty or misleading. Not referring to creativity would be as irrelevant for the word *like* as applying Langacker's archetype of billiard ball movements suggested by Jackendoff's formal characterisation Beethoven as the liker in the relation of liking and Schubert as the thing liked. This is completely and even fundamentally wrong.[5]

The next anecdote suggests the reason why Beethoven might have used the words "divine spark". The answer is not very difficult. A few years earlier Beethoven had composed the ninth symphony whose last movement contains the musical setting of Schiller's Ode to Joy. As is well known the first verse refers to the divine spark. The gist of this anecdote is that it brings us near to aspects of creativity that should now be studied, namely the *creative interdependency of language and music*. I hope that you allow some more details that, after all, indicate processes of mind/brain integration not yet mentioned in Jackendoff's project, though Jackendoff is a linguist who is also experienced in music. He will know that, at the epoch, there was a break in composing phrases or sentences using parts of the meanings expressed by its words, a break from Vienna classics to Romantics. The musical spirit in which Beethoven composes in Schiller's Ode to Joy is indeed quite different from the musical romantic spirit in Schubert's famous compositions. Perhaps Beethoven felt that a new

[4] Though in R. Jackendoff (1983) p. 27–28 the author is quite clear that the underlying framework or archetype implies aspects of a musician's creativity that are involved in characterizing the meaning of the sentence.

[5] Jackendoff remarks in a footnote that persons must be categorized as dot objects of the type object dot mind. It must, however be mentioned that the different mental, intellectual and emotional foundations of *like* are not implied in the notion of dot mind.

era of musical style was coming and Schubert presented it perfectly. Beethoven had just composed the Ode whose first verse was "Joy, beautiful spark of Gods!" This may explain his remark about Schubert.

The ambiguity of the word *compose* in grammar and in music, and even more so when *composing words and sentences,* means to compose songs or a mass is an important indicator of kinds of the brain's creativity linking language forms with musical forms. There is a musicologist[6] that explains in which way the German word stress in sentences was particularly influential in German composers.

The previous analysis of the example sentence demonstrated that it is completely inappropriate to explain the semantics of the feeling-based verb *like* by characterizing it only in some formal aspect, namely as the potential centre of a sentence of "Event"-type. In my view even this formal aspect relying on sentential argument structure is problematic. It would be appropriate for a sentence expressing object relation, as our example *The little star is beside a big star* but inappropriate for a sentence like *Beethoven likes Schubert* in which the subject, namely Beethoven, contains the state or attitude expressed by the predicate, namely by *liking Schubert.* In this sentence *Schubert* might well be better understood as a predicative[7] than as an object of the sentence. After the clarifications required by the example sentences I will now return to our main topic, namely language in the brain, and the challenge of explaining integration in the mind/brain/body system.

7.4 The fourth stage of pushing the world into the mind/brain/body

The core problem of a fourth stage, which I consider as a necessary modification of Jackendoff's third model of the "pushing the world" project, is an appropriate understanding of the integration of quite a number of components in each mind/body/brain individual. I will first compare functional aspects of *Jackendoff's last models* with my proposal for an extended modification that still correlates its principled organization with Jackendoff's models. Their correspondence in systems of functional neurocognition will be presented later.

[6] Thr. Georgiades (1982) *Music and Language*. Cambridge: Cambridge University Press. The book contains a chapter about German Language and Music.

[7] O. Jespersen (1924). *The Philosophy of Grammar*. London: Allen and Unwin. pp. 158–159.

My modified model will be presented in Figure 7.3. It should be compared with Figure 7.2, which is a slightly modified copy of Jackendoff's conceptualist core schema, presented above by Figure 5.2 (Jackendoff 2002, Figure 10.5 on p. 305). The modification of Jackendoff's model is merely topographic in folding upward Jackendoff's lower combination of Integration and f-knowledge base, presenting now the top areas of the f-mind/brain. The schema indications of the organizations of language, concepts (combined with inferences), perception and action, and finally the external input receptions of noises and concrete object signals are below. An additional remark is interesting with respect to the area of concepts. Recall the discussion of the sentence *The little star's beside a big star* and its interpretation on the level of spatial structure (SpS). Jackendoff's theoretical explanation for this level resulted, as discussed in section 7.2, from his geometry-based conceptual account. Consequently the area of concepts in our figure might now be interpreted to include both the conceptual system (CS) and the spatial system (SpS).

In my modified model presented in Figure 7.3, *integrations* mark the dominant characteristics. In view of neurocognitive analyses the various levels of the CS and SpS are accessible by pre-frontal cortex integration. The second range of pre-frontal integration may be understood as organizing selections among alternatives in various conceptual knowledge frameworks, theoretical, practical or imaginative. The next lower level consists of automatically processing

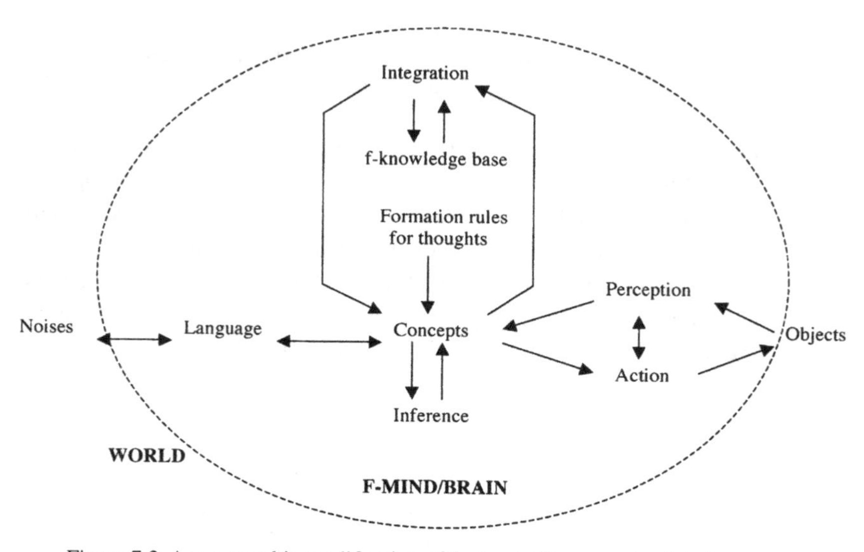

Figure 7.2. A topographic modification of Jackendoff's conceptualist view.

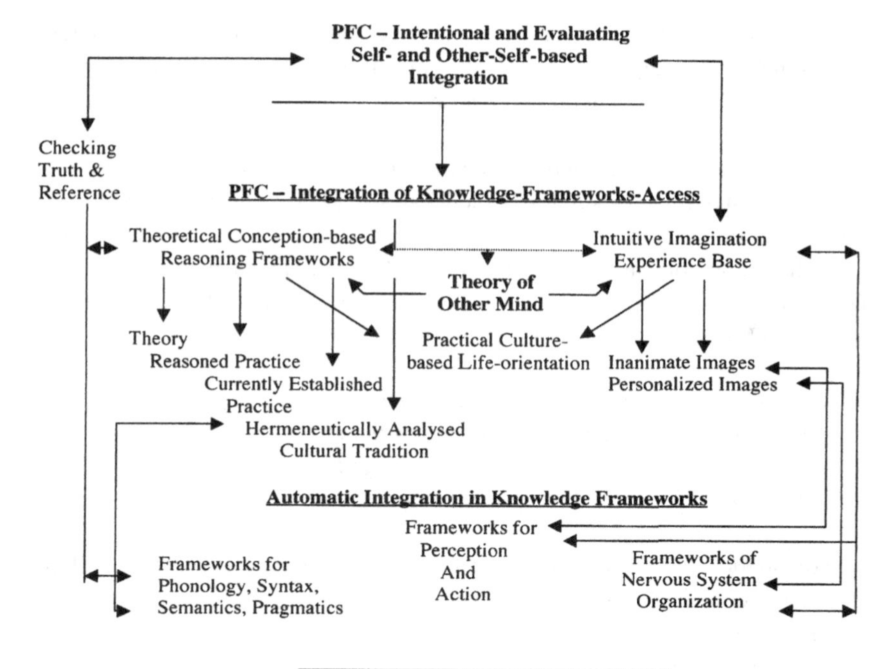

Figure 7.3. Integration of knowledge bases and knowledge dynamic.

knowledge systems that, given common sense or fluent usage experience, may independently operate automatically, that is without access of the pre-frontal selection. The processes may be located completely in the perception–action cycles (including organizations of language form and formal meaning) or in the autonomous nervous system (in the case of self-feeling). Saying that these processes are automatic means only that they can be but do not have to be autonomous in their operation.

Comparing my schema for the modified system in Figure 7.3 with Jackendoff's structure in Figure 7.2 it becomes obvious that similar to most other linguists, Jackendoff concentrates preferably on the left side of Figure 7.3. Let me add two further remarks. The area of *hermeneutics* is to be taken in the wide sense. It should not only include historical hermeneutics but also anthropology, ethnology and comparative linguistics whose reflections are partially based on systematic comparative linguistics. Many studies of Anna Wierzbicka (1999) about emotions and cultures are of this type. The importance of the area of the theory of mind requires more detailed discussion in the following sections. The connection between the abstract theory on the left and the intuitive imagination schematizes what was already discussed in references to Einstein's explanations, above in section 7.1 and in Chapter 4, section 4.1.

I think some more detailed explanations of the systems may contribute to better understanding. Jackendoff's model in Figure 7.2 presents *three layers* comprising integration with knowledge base, formation rules with three conceptualized components (language, concepts (CS+SpS) and perception–action). The latter is connected with the layer of inferences attached to the concept system. In Figure 7.3 my model presents four levels

(a) *PF-Intentional Self integration* with annex *checking truth and reference*,
(b) *PF-Knowledge framework access integration*,
(c) *Automatic integration in dynamic knowledge frameworks*,
(d) *Innate pre-cultural proto-self integration*.

This model is motivated by the intention to focus mental functions. They are mainly characterized by the following competences of persons: The characteristics of own *self*, imaginations of *other selves*, and *each-other self feelings* as well as their *ways of creativity*.

These functions obviously provide semantic aspects that clearly differ from the externally oriented perception–action organization, or the internal body oriented components such as the guts etc. and the intermediate emotion and feeling organization activated in the limbic system. The external or internal identifications of objects or situations, usually understood as the classical functions of objective fact-concerned cognition, as well as truth and reference semantics, contrast with emotion and feeling states of persons and probably also with "feelings" in higher mammals. In this functional perspective the meaning concentrates primarily on personal or individual core-self experience evaluation states – good or bad, or perhaps neutral. Their primary orientation is not the assignment of content, intention, truth or external or internal object characteristics. Instead what counts is rather its *value for the core-self's life* and *intellectual or emotional evaluations*. In these cases the core self is not the intentional agent of action or perception and is not very interested in checking the truth of what happens.

Language provides specific meaning distinctions for these aspects. The word *meaningful* has five entries in the Collins Cobuild English Language Dictionary (1987). Entries 4 and 5 are relevant about this term

4. A meaningful relationship, experience, discussion etc. is serious and important in some way, especially emotionally and intellectually.

5. If your life is meaningful, it has a purpose and is worthwhile.

Relevant are also some meanings of the word *relevant*:

1. If something is relevant it is connected with what is being talked or written about.
2. If something is relevant it has an important connection with something else, for example, with something of relevance to your way of life.

Core terms are *relationship* and *connection* that are emotionally or intellectually evaluated. Our notion of integration is exactly what these meanings of *meaning* are about. In our next chapter sections 8.2 and 8.8 will further explain the influence of usage components (background, prototypes, dynamic conditions of narratives as well as the requirement integration for model constructions).

Let me return to further discussion of the four layers of my integration model, from bottom to top. The classical system elements on the layers 2 and 4 were explained above in the Chapters 2 and 4. I should merely emphasize again that the autonomous and automatic organizations are based on innate global neurocognitive network structures. They are adapted by fine-tuning of distinctive features and classificatory structures to the external and internal environment, relative to the cortex. This fine-tuning of automatic organization occurs during learning and acquisition periods of childhood and adolescence. In later periods the development and role of the pre-frontal cortex develops also and improves its attention function and selectivity. Let me emphasize in particular Fuster's statement that the complete network of the automatic systems are best understood as a complete system of knowledge open for both *self-regulating organization or as memory system on which the pre-frontal cortex can work*. Due to its own dynamic character, it should not be called working memory. The fact is rather that the pre-frontal cortex, and even the automatic systems themselves, work with the *network memory for short times*. The components that are less clear on layers 2 and 4 of our schema are those of the right side, organizing inanimate and personalized images under intuitive imagination and image structures generated in the experience base, and in many cases the framework of nervous system organization, for instance activated in fitness training, jogging, but also meditations as practised in far-eastern meditation techniques, in many cases probably also in practices of meditative prayer, in particular in ritual contexts, and finally listening to meditative music or concentrated interpretation of impressive paintings or sculptures. All of these cases strongly involve the body and body-felt emotions and feelings. These receptive procedures are probably closely related to intuitive creativity in music and art, as will be discussed in the next chapter. Less known are also the two other levels, the innate pre-cultural proto-self experience and the intentional or evaluation-based self-integration. But what is important in the present context was already sufficiently discussed elsewhere in other chapters of the book.

What about the *highest level*? I feel that linguists usually focus on appropriate explanations that are indicated on the left side of the model. In a well-founded phenomenological way Searle (1998) provided fundamental explanations. It should now be sufficient to mention two aspects, *intentionality* and checking the *satisfaction condition*. On the left side of the model the phenomenology corresponds to neurocognitive aspects of attention operation, as for instance in visual perception of object and situation determination (see. for instance section 5 of Chapter 3). Contrary to conceptually constructive attention, intention is primarily dynamically active relative to its counterpart, the action content or the perception content. Selective access to knowledge is indeed similar. In addition there is another aspect to be organized. The intention content is somehow the intention's object on the mental level. But in the usual perception or action execution an error of intentional understanding cannot be excluded. But the organism has internal methods to determine the probability of truth, which is the degree of satisfaction of the intention. There are neurocognitive checking procedures, either by concentrating the focusing relevant for the perception or action event, or checking the compatibility of the object or situation with the environment's data, or finally by asking reliable partners about the given situation's probability of truth. Thus various cooperations with other parts of cortical operations can serve the *determination of intentional satisfaction*.

The situation is different for the *right side of the model* on which the criterion is less one of rigorous identification and truth but rather one of intuition. Experienced mathematicians and theoretical scientists, for instance Einstein and Gödel, emphasize that on their way to new ideas or solutions of problems they strongly rely on imagination. Due to their experience their intuitions lead them to probable solutions. In this situation they start checking their ideas by applying formal proof procedures. *Creativity of imagination and control procedures of formal proof* cooperate in finding true ideas. The scientist's first step of creativity may be similar to those relevant for musicians and artists. However, the second step is different. The mathematician's and the scientist's checking are based on formal control procedures leading to what is accepted as a proof. Certainly the artist also checks his criteria but the criteria for judging the artistic quality are usually less reliable.

7.5 An important archetype: togetherness and each-otherness

In Chapter 5 the discussion of a basic problem for Jackendoff's project of "pushing the world into the mind/brain" was introduced by the section's headline "The world of thinking and knowing, loving or hating, happy or sad mind/brain/bodies." I argued that, in pursuing the project, it is necessary to acknowledge that each of the three perspectives with focus on the mind, on the brain and on the body refers to existence, that is involves continuous activity. Dynamic mind/brain networks activate thinking, feeling, and speaking in the cortex. Bodies have, as we learned from Cannon, "their own wisdom" organized by the autonomic nervous system. All three, mind, brain and body are *living entities* with inner dynamic. I also argued that a "world pushed in the mind/brain" must contain, in its system of mind/brain's attention, activities referring in a specific way to other mind/brains. As already mentioned elsewhere thought in mind/brain can mentalize or empathize, that is to say can place itself into the position of another mind/brain's thought.

This capacity to pay attention to other mind/brains must also be conceded to the other person. Thus mentalizing is potentially reciprocal. Since Premack and Woodruff (1978) there has been much research about these phenomena. The consequence is that Jackendoff's (2002) project cannot be satisfied with "pushing the world into a single mind/brain/body" but must *introduce a plurality of mind/brain/bodies* who potentially *pay attention to each other*. Those that actually are often in the situation of paying *attention to each other* are thereby in a *group of mind/brains* that are in this sense *together*. The group forms even a kind of unit, a *social unit*.

Before discussing in which way each-other attentions and actions constitute social units I will first continue and finish the discussion of Figure 7.3. The horizontal line and the two other lines, connected with the Theory of Other Mind, indicate the interdependency of relevant mental functions. They connect the Theory of Other Mind area with theoretical areas on the one side and the imagination area on the other side. In a sense it also includes the creativity mentioned by Einstein, since it relates both areas of thought, the formally theoretic ones with imaginative ones. Note that the long vertical connection, which indicates the fundamental connection of the cortex – the "intelligent wisdom"– and the nervous system area – Cannon's "body wisdom"– discussed in previous chapters, passes by the imagination area, connected merely by pointed lines. There the intention of the pointed line is to indicate that *thought is involved* and not the concrete combination of concrete attention and consciousness of combined emotion, feeling and pre-frontally integrating experience and activity. Instead of *concrete experience-based self*-knowledge and *self*-feeling combination, as they were discussed in connection with Damasio's proposals, the theory of mind area generates a sympathizing thought referring to the other person. The relevant imaging involved relies on sympathizing feeling and not on concrete feeling. Consequently, we must emphasize the difference between the experiences of *types of self* discussed by Damasio (1999) and the *acts of sympathizing* (the mind/brain's putting itself in the place of another mind/brain) based on imagining and thought.

Often there are certain errors when words referring to feelings and emotions are involved. There is an interesting controversy about emotion and feeling in which Wierzbicka (1999) criticizes two authors who she cites in her introductory section: "Emotions are thoughts somehow 'felt' in flushes, pulses, 'movements' of our livers, minds, hearts, stomachs, skin. They are embodied thoughts, thoughts seeped with the apprehension that 'I am involved' ". The next author comments "This apprehension, then, is clearly not simply a cognition, judgement, or model, but is as bodily, as felt, as the stab of a pin or the stroke of a feather." Wierzbicka is not quite satisfied. She agrees with both authors that some thoughts are linked with feelings and with bodily events, and that in all cultures people are aware of such links and interested in them (to a varying degree). Now comes her critique: She does not agree that "feelings" equals "bodily feelings". For example, if one says one feels "abandoned" or "lost", one is referring to a feeling without referring to anything that happens in the body. Precisely for that reason, one would normally not call such feelings "emotions", because the English word *emotions* requires a combination of all three elements (thoughts, feelings and bodily events/processes).

Obviously, Wierzbicka needs some clarification. In a sense the brain is a part of the body and any thought has a corresponding process in the brain, partly in the cortex, partly in sub-cortical areas and partly in the body, for instance by having no dynamic of muscle movement etc. Consequently when feeling abandoned or feeling lost something happens in the brain and in parts of the body. What Wierzbicka apparently wants to emphasize is that in these two cases there are no visceral components – thus body's core elements – that *cause* uneasiness or pain. Agreeing with another author she writes that the very meanings of words such as *shame*, *anger* or *sadness* on the one hand, and *hunger* or *thirst* on the other draw a distinction between feelings based on thoughts and purely bodily feelings. But the point is not that one class of word meanings expresses feeling as a *kind of thought* and the other is emotion as a *kind of bodily event*. We should rather consider *causes* in distributing activity of brain areas – brain in the wide sense – and processes of distributing caused activities in the brain. Anger may be caused by evaluations of the pre-frontal brain leading to consequences in the bodily emotions and movements and hunger is rather caused by visceral phenomena influencing the hypothalamus and subsequently generating negative feelings and thoughts. Distinctions are based on the mind/brain/body's changes of processes, not of this or that putative restricted type of functional source: body or thought. After all thought is also organized in the brain.

7.6 Social groups, institutions and each-other relations

In the previous section I mentioned the fact that those that actually are often in the situation of paying *attention to each other* are thereby in a *group of mind/brains* that are in this sense *together*. It appears that this situation is normally differentiated into two directions, friendship and sympathy, as in the case of friends, married couples and other pairs of mutual sympathy, and mutually vivid attention as practiced in a group or another kind of unit, in any case a *social unit*. When Searle (1998) addressed this problem he was motivated by the philosophical challenge to explain how there can be an objective social reality that is partly constituted by an ontologically subjective set of attitudes. Two most important interrelated features, consciousness and intentionality, core features of mind, determine his framework. Searle insists that these mental phenomena are essentially a biological phenomenon, a view that I understand as being similar to Leibniz' organismic framework as presented in section 2.1.

Starting to approach the social reality Searle proposes the following move with which I sympathize (Searle 1998, p. 118): Switching from "I intend, believe, hope and so on" to "We intend, believe, hope and so on" the "we-statement" implies an "I-statement" for each participant of the group. "We intend x" implies "I intend to do my part in the distributed activity of the members of the group." Searle still believes that there is an irreducible class of intentionality or "we-intentionality". Shouldn't it be possible to reduce the "we-intentionality" into individual intentionality? Searle's strategy is to define arbitrarily a *social fact* as any fact involving two or more agents to have collective intentionality (p. 121 top) and he adds his earlier notion of a constitutive rule. The common knowledge of the community relevant classes of rules that each social group member has acquired and normally and regularly applies in his behaviour provides the guarantee of the proper existence of the social group.

Searle seems to believe that his argument solves the problem without reduction to individual intentionality of behaviour. The reference to a *principled rule orientation shared* by the members of a group seems to be sufficient for eliminating any individual reference frame. This foundation seems to be correct for social units or communities that aim at stable and appropriate distribution of behaviour roles in the group. Properly defined the reasonable structure constitutive rules are similar to laws in natural science.

Searle's socially shared rules have a counterpart in Jackendoff's (2002) proposal how to escape the accusation of solipsism. He acknowledges that "by pushing the world into the mind, we no longer have a standpoint in objective 'actuality', independent of the minds [of speakers]". He exemplifies this as follows: "Joe says: 'Look at that duck!' But in actuality he points to a platypus." The reference to actuality relies on an objective world that no longer exists in the model after it was pushed into the mind/brain. In a first move Jackendoff seems to accept not only one ideal mind/brain but as many as there are observers. This would allow saying "In the conceptualized *worlds of other observers Joe* points to a platypus" while Joe believes himself to be certain of seeing the duck. Jackendoff feels confronted with the question "What is to stop them from all having different conceptualizations? How can we speak any more of Joe's being mistaken? And why should we care? Are we therefore doomed to a radical relativism about knowledge?"

Jackendoff's first answer seems to be plausible: A way out comes from recognizing the sound sequence

Joe says "look at that duck" but in actuality Joe points to a platypus

not as disembodied pieces of language "out there in the world" but as originating from a speaker who is communicating his or her conceptualization of the world. This conceptualization includes two things: first the judgement [in the mind/brain] that Joe said something with intent to refer to some entity, and second that this entity was a platypus [in the mind/brain]. And now we can ask how the speaker [the mind/brain] comes to this judgement.

The *basic question* is "How can [in the mind|brain] someone rely on someone else's report of an observation, and how can [in the mind/brain] anyone count on others' usage of language. Turned in a different way: if conceptualization [in the mind/brain] is essentially personal [how else could the mind/brain be?] how can we communicate? Now in this way Jackendoff's notion of a unique mind/brain becomes confused. Who [in the mind/brain] is someone else, or an other? Who are we? After pushing the world into the mind/brain did we also push all people in the world? But would communicating remain merely as a concept in the set of all concepts in the unique mind/brain?

Jackendoff seems to feel the problem when he writes "This is now getting quite speculative, but I am inclined to think that human beings have a need to 'tune' their conceptualizations to those of others – to have a common understanding of the world." He now seems to understand that there cannot be a single and unique mind/brain as there is, in common sense understanding, a single and unique world. There must be at least as many mind/brains as there are human beings, or more generally beings that can communicate meaningfully. It is only then that the mind/brains can tune their conceptualizations.

After this move we can agree with Jackendoff's following statements: "Since we cannot read minds, the only way we can assess our attunement is by judging whether the behaviour of others, including what they say, makes sense. To the extent that all members of a community are effective in acting on this drive, there will be a tendency for conceptual convergence, including in the use of language." But there is one point that must be added: If we further analyse this move, we must also consider the possibilities studied in theories of mind, and of the practice of mentalizing, sympathizing, and placing oneself in somebody else's place or position. Concerning conceptualization we can no longer assume a single and unique knowledge system of all mind/brains but only acquired similarities to the degree that depends on intensity of each-other "tuning" in communication.

These arguments explain why I feel forced to introduce the fourth stage of Jackendoff's "pushing" project. It must necessarily be based on each-other communication. I think that it is wrong to consider only the basic role of accepted and conceptually established rule systems, or the basic role of conceptual systems in mind/brains. The neurocognitive perspective envisages a more complex and extended system in which the rational mind of the cortex interacts with body-based dynamic generating the equilibrium of awareness and feeling. It is the each-other practice of existing groups that provides the generative core of smaller or larger groups of people. Consider the models of two or a few members building a *group of friends*, or *of a family*. They may also have established a few constitutive rules, but the basic foundation does not rely on the fact that the members concentrate on following or obeying the rules. There are other characteristics that contribute to stabilize the group as a social unit. Here the *positive each-other acts and attitudes*, such as friendship or family solidarity are of primary importance. Looking for original criteria one may refer to partially innate and partially childhood-acquired evaluation feelings. Already from this stage onward they distinguish each-other actions and attitudes in their positive, negative and neutral values. I think that linguistic, psychological and neurocognitive studies of the range of each-other actions would be particularly fruitful. In view of initiating such studies the next section presents a list of each-other verbs. The list is certainly not complete but is instead at least supplemented by some words that belong to the group in meaning though not in form.

7.7 Each-other words as linguistic entrance to the lexical semantics of altruism

There is a clear contrast between the meanings of the noun *self* in English dictionaries and in the language of neurocognition.

If you want to know what the self is, some dictionaries give the following explanations. The Longman dictionary tells us that the self is the whole being of a person, taking into account their nature, character, abilities, etc. One may perhaps think that an explicit biography may characterize the person's self. Collins gives a similar presentation in writing that your self is your basic personality or nature considered especially in terms of what you are really like as a person or what you are really like at a particular time in your life.

Additions of adjectives may seem to specify more common meaning indicating distinct feelings relative to one's normal and usually not very conscious feeling of self experience. Thus the Longman dictionary gives the example: "I am feeling better but I am still not quite *my old* self (= as I was before illness)" and Collin's example emphasizes "He was his *usual imperturbable* self." Or "She was *her normal* self *again*." Or even "Once dressed, they became their own decisive selves again." One of Langenscheidt-Collins examples was "He showed *his true* self," expressing a contrast to what could have been expected before.

All of these examples differ in their meaning from Damasio's use of the word *self*. Knowing this, Damasio explains that his three neural-self notions, namely proto-self, core self, and autobiographic self are neurocognitive experiences emerging at different stages of an individual's development. In our present discussion only notions involving conscious experience of adults are relevant, that is either core-self or autobiographic-self or mentalized selves of others.

As already discussed in section 4.6 the aspect of self-feeling is a meaning component of the first person pronoun that is also mentalized in thoughts about other persons. This contrasts with a *common linguistic theorizing* according to which the person pronouns have primarily *deictic character*. If somebody expresses feeling bad deixis is only a secondary meaning of the pronoun, the primary meaning being the bearer of the feeling, namely the self of the speaker. Consider the emotions and feelings caused by strong stomach pain in a baby and in an adult. The expressed meaning is the same whether crying or using a pronoun. Hence both have the self as a base of feeling experience. The child's self indicates already at the origin an experiential continuity. It *later becomes* a meaning component of the first person pronoun as soon as the child learns to use its feeling-based self reference as its own marker in deictic contexts. The meaning component can also be mentalized for second person pronouns or third person pronouns when they refer to living beings. I think that these arguments are sufficient to use the neural notion of self in meaning definitions for each-other archetypes.

Before concentrating on this meaning a short account of grammar may be useful. Grammar assigns the verbs with each-other object to the class of *reciprocal verbs*. A grammar may characterize the property of reciprocal verbs by the following meaning relation: If we know that "John met Mary" we also know that "Mary met John" and that "John and Mary met" or also "John and Mary met each other". Obviously the names in the last two sentences can also be inverted. The meaning relations would also be true when the names are replaced by more complicated noun-phrases that are NP_1 and NP_2. In a slightly formal way we could define the grammatical rule (with V as a variable for a reciprocal verb):

If "NP_1 V NP_2" then also "NP_2 V NP_1" and also "NP_1 and NP_2 V each other."

The grammatical object, namely the phrase *each other*, emphasizes the reciprocal meaning. In the last sentences the conjunctions of John and Mary can also be replaced by a noun-phrase in plural, say "The children" such that the following rule holds:

If "The children met." then also "The children met each other." or also
If "They met" then also "They met each other."

It should be clear that these rules apply only to verbs that are classified as reciprocal or each-other verbs. There are some reciprocal verbs in which the object position of each other may be combined with a preposition. Take for instance the verb *argue*. Two situations may be expressed as follows:

We were arguing about a small amount of money....
Should we argue with each other about such minor things?

In order to familiarize us with the use of the explicit reference to the feeling self we use the reciprocal verb *like*:

If "My self likes your self" and "Your self likes my self" then "Our selves like each other".

It should be clear what the statements want to express. It is not *I* or *you* in its deictic meaning that is involved but rather the neural combination of the feeling-emotion combination organized in the combination of the cortex with the nervous system. The positive, negative, or neutral evaluation of the verb meanings are organized in the emotion and feeling relevant brain areas. The semantics assigned to the sound-patterns of the verbs in the left column of Table 7.1 activate per se emotionally *positive signalling*, the verbs of the right column *negative* and the verbs in the middle *emotionally neutral* ones.

It may be suggestive to *illustrate roles* of some of the verbs in the three columns.

Here are some examples for the positive column:

Positive personal feeling attitude, some of them indicating developmental stages:
- finding each other attractive,
- liking each other,
- looking at each other lovingly,
- judging each other with great sympathy and esteem,
- cheering up each other.
- falling in love with each other,
- loving each other.

In the Jewish and Christian religions in which each-other-love plays a central role, the words for love have a central and more fundamental status relating persons, their personal souls, emotion/feeling and the life-relevant mental knowledge framework.[8] The following statements distinguish three characteristic stages[9]:

- Mutual mental *possession* of the souls: I am yours and you are mine.
- Being mentally *in each-other*: "I am *in her/him* and she/he is *in me.*

[8] Compare Deuteronomy 6.5: "You shall love HASHEM your God. With all your heart, with all your soul and with all your resources." Cf. Tanach (The Stone Edition 1998) Lv 19.18. "You shall *love* your fellow as your self." Also Tanach (The Stone Edition) p. 293. Here you find also the following paraphrases using attitude words, (a) by Hillel: "*What is hate*ful to you, *do not do* to others" and (b) a commentary by Ramban: " … we must treat others with utmost *respect* and *consideration.*" The New testament, Matth. 22. 37–40: "Thou shalt *love the Lord thy God*, with all thy heart, and with all thy soul, and with all thy mind. This is the first and great commandment. And the second is like onto it. Thou shalt *love thy neighbour* as thyself. On these two commandments hang all the law and the prophets." (Authorised version of the King James Bible.)

[9] Compare the following examples: (a) "I am *my beloved's* and my beloved is *mine*." (Song of Songs 6.3; lit. cf. Tanach 1998, p. 1695.) Compare also Bach's cantata BWV 172 The Altus part in the Duet Aria; (Commentary Tatlow Interpretation text: "Expressed in the language of lovers, the impatient desire of the believing soul (the Duets soprano) contrasts with the tenderness of the kisses of the Holy Spirit (countertenor). The emotional climax of the duet coincides with the union of the soul, making a threefold invitation 'komm herein' ('come in!') and the Holy Spirit singing 'Ich bin dein und du bist mein' ('I am yours and you are mine')". "(b) Reference to John 14.11 "I am in the Father and the Father in me." John 14. 20: "At that time ye shall know that I am *in* my Father , and ye *in* me , and I *in* you." 14. 21: "He that hath my commandments, and keepeth them, he it is that *loveth* me: And he that loveth me shall *be loved* of my Father, and I will *love* him, and will manifest myself to him."

Table 7.1. *A list of each-other verbs*

Positive	Neutral	Negative
accompany	acclaim	abandon
admire	approve	accuse
adore	call	beat
affect	congratulate	bite
appreciate	communicate	blame
assist	compete with	bore
attract	elect	criticize
bless	examine	deceive
believe	excuse	demoralize
calm	expect	despise
care for	face	disappoint
charm	know	fight
encourage	greet	force
fascinate	identify	frighten
forgive	imitate	hit
hold	invite	hurt
hug	look at	irritate
idealize	maltreat	kill
kiss	match	manipulate
like	meet	misinform
long for	mimic	mislead
love	mirror	mistrust
marry	motivate	offend
miss	name	oppose
pardon	obey	persecute
pine for	observe	ruin
promise	apologize	scare
protect	perceive	seduce
recognize	persuade	shoot
recommend	play with	shove
reconcile	pinch	struggle with
respect	quote	
safe	react to	
shield	remember	
stick to	remind	
strengthen	see	
sympathize with	speak to	
trust	talk to	
understand	teach	
welcome	thank	
yearn for	think about	
	wash	
	wish	

- Being a *unit*: She/he and I are one
- Being mentally *identical*: "The Father and I we are (absolutely) one.

By contrast, here are some contrasting stages of the deception words from the column of "bad words":

- disappoint each other's feeling of expectation
- disappoint each other's affection
- dash each other's spirit
- deceive each other
- cheat each other

It should be clear that more detailed analyses of semantics would require reference to phenomenological psychology and to neurocognition. Descriptions of the developmental psychologist Daniel Stern with implicit indications to neurocognition are particularly suggestive for a self oriented semantic that must necessarily transcend the common semantics that is more conceptually or perception–action oriented. Stern correctly emphasizes that, as soon as we concentrate on yoking of perceptual and affective experiences rather than merely of combinations of perceptions and of perceptions and actions, we are concerned with less systematized experiential concepts. This is particularly true for hedonic tones in which associations of perceptions become affect-zimbued. In Freud's view affects not only make perceptions relevant by way of association, they also provide the ticket of admission for perceptions even to get into the mind (Stern 1985, pp. 64–65).

As Stern emphasizes these considerations are particularly important to clarifying the notion of self-with-other. Stern's explanation is quite clear: "The sense of being with another with whom we are interacting can be of the most forceful experiences of social life. Moreover, the sense of being with someone who is not actually present can be equally forceful. Absent persons can be felt as potent as almost palpable presences or as silent abstractions, known only by trace of evidence. In the mourning process, as Freud pointed out, the one who has died almost rematerializes as a presence in many different felt forms. Falling in love provides a different normal example. Lovers are not only preoccupied with one another. The loved other is often experienced as an almost continual presence, even an aura, that can change almost everything that one does – heighten ones perceptions of the world or reshape and refine one's very movement. How can experiences such as these be accounted for in the present framework?" (Stern 1985, p. 100).

I think that Stern's characterization of "Self-with other" indicates quite well the direction that semantics would have to take in future explanations of each-other relations. It is also quite clear, that we are confronted with yoking of different components of brain and nervous system, the cortical areas of perception and action as well as of cortical areas of conceptual thought and imaging creativity, the sub-cortical areas involved in emotion and feeling as well as of bodily feelings. Pre-frontal organizations will be involved as well as they usually are in other co-ordinated organizations. Certainly, Stern's explanations apply to larger areas of phenomena than the archetypal area of each-other verb semantics. But they apply also to our special cases. The fact that these phenomena are not yet well accessible to scientific analysis does not imply that their phenomenological explanations should not be taken into account.

7.8 The positive each-other perspective as a transcendental ideal

Langacker (2008) introduced the notion of an archetype as an interpretative domain in which interdependencies of words can be systematically accessible and brought to efficient, flexible and appropriate use. But our each-other archetype is not yet acknowledged as such. It is obvious that in this case the support of psychological and neurocognitive analyses might be helpful. Mere reference and aspects of truth in external world are concerned with completely different semantic facts. Coming originally from a more formalized direction of formal linguistic analysis than Jackendoff (2002) transcended, in section 5.4 of Chapter 5, the external world related logical perspective to a perspective of perceptual analysis. It is clear that we need more extended frameworks for the analyses of semantic structures.

In some cases, support from psychology and neurocognition is not enough. Often our linguistic studies may get help from philosophically founded analyses. At the same time the openness for scientific interdisciplinarity would certainly even better support an integration of scientific studies with philosophical frameworks. This does not exclude that some ideas bring us near to reflecting reasonable foundations for formal and empirically systematic foundations.

Here the philosopher Kant has given guidelines. I think that the notion of each-other relations should connect the personal world mind/brain/body units, which were introduced in our proposal for adding a fourth stage to Jackendoff's "pushing the world project." As a consequence the constitutive world consists of *interrelated mind/brain/body worlds*. This does not exclude that the mind/brain/bodies not only establish each-other communities but may have stabilizations, supported by *constitutive rule institutions* as Searle (1998) suggested. Both each-other-relation (Theory of Mind, mentalizing, sympathizing) and knowledge and agreement in the practice of constitutive rules in social units could contribute to establish possibilities to clarify aims for ideal guidelines toward improved world communities.

In considering "foundations of foundations" we should learn from Kant to use transcendental ideals in framing concrete principles. The set of positive each-other relations could hint at such a transcendental ideal.

Just as with Kant's transcendental refutation of Hume's arguments against causality, a deeper study of the each-other archetype might well serve as a refutation of Dennet's claim: " Darwin's idea is a universal solvent capable of cutting right to the heart of everything in sight" (Dennet 1995, p. 513) except when the notion of "sight" is taken in a very limited way that would exclude many cases of insight.

8

Dynamic language organization in stages of complexity

8.1 The gap between formalist structure definition and neural dynamic

Acknowledged mental knowledge theories and neuronal brain analysis should not remain separated. The challenge is particularly addressed to linguistics as a mental study both in the abstract form of formal linguistics and in formats of "Grammar as life". In the present research situations we cannot do more than propose and study models that seem to be plausible for bringing structures and organizations of the linguistic mind and the neurocognitive brain into correspondences. Even this is a difficult task, in particular when we want to correlate techniques of passive formal symbolic structure representations and central processing combination with systematic descriptions of connection-based and radically dynamic organization in cognitive neuroscience! This was already explained in earlier chapters: In principled biological perspective there are only self-organizing units in cognitive neuroscience in Aristotle's, Spinoza's and Leibniz' philosophy, today scientifically systematized by Fuster and Damasio.

We must expect that perspective correspondence is more difficult to model than the similar task of relating software and hardware processes in computer science. The "software" and "data structures" that appropriately constitute the functional mind must be far more tightly bound up with the biological nature of the neuronal networks. This certainly holds for the relation between mental linguistic structure and neuronal brain structure-based organization.

Most linguists are sceptic. Typical is Chomsky's argument that the gap between formalist linguistic theories and neuronal descriptions of language organization cannot be bridged (Chomsky 1963b. p. 326) claiming that psychologists have long realized that a description of what an organism does and a description of what it knows can be very different things. Linguists have understood that a formal study of grammar models is a mental approach to clarify the knowledge achieved by the mature speaker of a language. It is not concerned with the description of what a speaker or hearer does in his acts of speaking or understanding nor how brains organize these acts. However, it must be assumed that the neural organization somehow represents knowledge. The basic question remains: How are phenomenological mental experience or the formalized mental knowledge available in the neural brain. The sceptical stance seems to be justified: One can scarcely hope to develop a sensible theory of the actual organization of language before the basis of a serious and far-reaching practical analysis or formal account of what a language user knows has been completed.

I do not share this separatist view. This does not mean that I would not acknowledge practical and formal progress based in well-established frameworks and analyses of linguistic research. We should develop linguistic research in a spirit that defines challenges for neurocognitive modelling of approaches of various types of phenomenological and linguistic principles of organization. The first section will exemplify this in three phenomenological dynamic orientations.

But as emphasized several times already we should not merely expect better insight only from our home discipline. New perspectives will come from interdisciplinary inter-translations of perspectives. We should start right away with developing and studying stages of complexity in well-founded working models and their mutual translations This was, by the way, already Chomsky's solution, when he developed mathematical linguistics, in the 1960s. A similar initiative must study empirically based working models of neuronal organization. Fifty years ago Chomsky selected a simple type of formal model, the constituent structure grammar (Chomsky 1963a, p. 292). It was clearly understood as an initial framework for his studies, and is indeed a good choice for my studies also (Schnelle 1981, 1991). It seems to be useful to present exercises helping us to develop expertise in mind-brain correspondence modelling before we will be able to solve more complex and comprehensive solutions.

Just as with our intention, Chomsky's task was also translational. He wanted to relate mere collections of intuitive linguistic knowledge or empirically observed facts to strictly ordered and systematically explained system descriptions applying mathematical structure models. What he was looking for was a mathematically formalized grammar, which would define a mathematical theory of syntax providing the core component for a more comprehensive mathematical linguistics that would satisfy the challenges of the discipline. He knew that his pre-mathematical intuitions of a linguistically more comprehensive and more adequate form for a transformational grammar, were too difficult to be solved in a single swoop. As a mathematical form of grammar structure, he therefore concentrated on the formal understanding of *constituent structure grammars*. Although he explicitly acknowledged the short-comings of this type of grammars he wrote a complete chapter about formal properties of grammars in which he emphasized that there are very good reasons why the formal investigation of the theory of constituent structure grammars should be intensively pursued. Indeed, it does succeed in expressing certain core aspects of grammatical structure. I believe that trying to translate constituent structure grammars into neurocognitive models is even today a good start.

8.2 Dynamics in phenomenological and usage linguistic analyses without reference to the brain

8.2.1 Stern's mindscape and Searle's background

The observations of neuroscientists are clear. The specific constitution and structure of columnar modules is typical for primary sensory areas of the cortex. On higher levels of the perception–action system the distribution of neurons and local clusters of connectivity are much less regular. The basic fact is that core or "parent" neurons, namely the pyramidal neurons, are surrounded by cooperating excitatory and inhibitory inter-neurons that support specificity of the "parents" reaction. But the arrangement and the types of connectivity are less regular. The reason seems to be that the columns in the peripheral cortex are particularly important for determining specific distinctivity perception and production. Whereas distinctivity of movement behaviour generally develops in parallel the situation for language sound perception and articulation are different. Children have already a perceptive lexicon before they can speak during the first half of the second year (Karmiloff and Karmiloff-Smith 2002, p. 62). Learning articulation is later. It is also possible that, because Broca's areas are nearer to the ventro-lateral pre-frontal cortex, during development they required support from emotion areas and areas that identify things in the world .

But later in language acquisition precision of perception and articulation is important in many situations. The phyletic levels in the perception–action system serve just these tasks.

The tasks to be served by the higher levels in late childhood, adolescence and adulthood are very different. Stern's (1990) characterization is very impressive. What is developed is an *inter-subjective mindscape* that contains intentions, desires, feelings, attention, thoughts, and memories, all of those vivid events that occur in an individual's mind but are invisible to others.

Searle supports this psychological view in terms of philosophical phenomenology. He writes that in addition to quite a lot of specialized knowledge "I have to have a set of capacities, abilities, tendencies, habits, dispositions, taken for granted presuppositions, and 'know-how' generally that I have been calling the 'Background,'" Searle's general thesis of the Background is that all of our intentional states, all of our particular beliefs, hopes, fears, and so on, only function in the way they do – that is they only determine their conditions of satisfaction – "against a background of know-how that enables me to cope with the world" (Searle 1998, p. 107–108).

It is obvious that in Searle's more recent interpretation speech acts are not only determined by the specific intention and content of sentences. The proper selection of these core aspects requires access to the broader range of background knowledge. Given the dominant terms used by Searle, namely "capacities, abilities, tendencies", which he conceptualizes as "presuppositions" of the selected word or sentence expressions, the primary pragmatic status of his notion of background is clear. Reflections about how to relate these pragmatic characteristics may be related to the typical distributions of knowledge processes in the brain suggesting that *large parts of the perception–action system relevant for semantic processing* should be involved. Still more important is probably the conscious access of *pre-frontal selectivity* in the range of normally automatically applied knowledge. Selectivity is not a result of precise distinctions in the perception and actions periphery, such as those providing sufficiently clear articulation and understanding. It rather depends on widespread evaluation and selection in the ranges of neural knowledge memory. This corresponds well to the lack of local columnar modules of the periphery and the distant organizations of selectivity. The general principle is clear but the detailed description of specific neurocognitive networks is very difficult, given the present status of research.

The situation is not much better when our studies would concentrate on the linguistic core, namely on *morpho-syntactic and* semantic *categorization of words and sentences*. The *structure-based* studies of *Sgall, Hajičová*, to be discussed in a later sub-section, will make this quite clear. But it is also important to consider *variable aspects of usage-based linguistics*. Various types of analyses have been proposed. Some statements of *Lakoff* are particularly well focused. In some respect they are similar to *Langacker*'s views, which I discussed in Chapter 6; the notion of flexibility is particularly important.

8.2.2 Lakoff's prototypes in flexible language use

Let me briefly review Lakoff's interesting core proposals. His basic idea is that *categories* are part of our experience as neurally organized beings. The categories we form are part of our experience, a statement that is very similar to Fuster's view based on externally oriented perception–action memory in the cortex. Lakoff agrees. But when he claims that this view contradicts meditative traditions (Lakoff and Johnson 1999, p. 19), he does not understand that focusing *on perception–action procedures* does not at all exclude *procedures involving the autonomic nervous system* which allows breathing during concentrated thought. Most meditative traditions, for instance Pranayama,[1] involve long-practised pre-frontal concentration on body-related processes of the somatic muscle systems, thus not involving the externally oriented perception–action system of the cortex. But this remark was merely incidental. Let us return to the main point.

Lakoff understands concepts as neural structures that allow us to mentally characterize our categories and reason about them. Human categories are typically conceptualized in more than one way, in terms of what are called *prototypes*. Each prototype is a neural structure that permits us to do some sort of inferential or imaginative task relative to a category. Lakoff is somewhat optimistic in assigning a neural structure. But his optimism seems to be based on the possibility that neural structures and their processes can be postulated and characterized on linguistic knowledge about categories and concepts only. But in fact his descriptions are merely directed to linguistic analyses of behaviour situations. This holds in particular for the following statements.

He distinguishes *typical-case prototypes* from *ideal-case type prototypes. The former* are used in drawing inferences about category members *in absence of any special contextual information. The latter* allow us to evaluate category members *relative to some conceptual standard.* Consider as an example the difference between typical husbands and ideal husbands. Since there are still other prototypes used for making probability judgements, for instance social stereotypes and salient exemplars, we may say that prototype-based reasoning constitutes a large proportion of the actual reasoning that we do. Reasoning with prototypes is so common that it is inconceivable that we could function for long without it. The use of prototype references allows us to simplify verbal arguments, thus somehow neglecting what we would know when clearly considering our background knowledge in Searle's sense, as characterized in the previous paragraphs. They characterize a move from explicit considerations of pragmatically available knowledge details to a usage linguistic-centred knowledge framework.

[1] B.K.S. Iyengar (1981) Light on Pranayama. London: Allen and Unwin as well as W. **Singer et al (2009)**.

The framework provides even more tools for flexible expression such as grading prototypical words. All of these means allow us to conceptualize categories in communicatively efficient ways. Less communicative and more formal or bureaucratic acts insist on strict standardized conceptual order, thereby however losing the flexibility of conceptual prototype expressions, which use the graded structures of categories and the, often useful, fuzziness of category boundaries.

From the neurocognitive point of view it is interesting that not only hierarchies of order are brought into play but also conditioned modifications and evaluations and their communicative appropriateness in view of fluency and flexibility instead of rigorous precision of truth and reference. But as already mentioned in Chapter 6 this flexibility and practical efficiency has a price to be paid by the neural network: Rigorous perception–action hierarchy as described by Fuster must be supported by circumstantially modifying mechanisms. When pyramidal cells organize the basically appropriate connections with other areas the inter-neurons of their environment support and control the contextually appropriate automatic varieties in the hierarchical orders. This property of pyramidal neurons and inter-neurons may well require a "neural processing neighbourhood" that must be rather different from the columnar modules.

8.2.3 The Prague School's communicative discourse dynamic

Up to now we have considered the influence of *mindscape and Background* knowledge and the use of *prototype flexibility* of words and expression. The conclusion was that their use in common sense statements and in discourse might be supported by degrees of variation and evaluations of situation-acceptable fuzziness. It is useful to study still other aspects of speech variations that indicate new challenges. The foundation of the new range of challenges derives from the fact that different types of languages rely on different types of morpho-syntactic organization. In particular the differences of morpho-syntactic organization often have fundamental consequences for the regularity of word order that the language type develops for common discourse and narration (Sgall, 2006).

Typical influence results from the ways in which *basic lexical word types* like nouns, verbs, ad-nouns (that is adjectives), adverbs, and words expressing typical spatial and temporal meaning relation, usually called pronouns are used. There is some reason that a differentiated analysis must come back to the classical opposition between *syntax* and *morphemics.* Similarly to *grammar* and *lexicon,* this is a pair of old concepts, richly discussed and shown as a cornerstone of plausible hypotheses. Both these oppositions have served for many centuries of linguistic research, giving ground for a modular understanding of language, also their boundary lines are blurred by "grey zones" of transitions, of intermediate phenomena between morphemics and word formation, syntax and analytic morphemics, in idiomatics, and so on.

There is the old perspective according to which languages with typical inflexion of basic lexeme categories were considered to be natural. In its relation to syntax, inflexion morphology is in any case more natural than agglutinating or analytic types of language:

(i) Each word form indicates the syntactic function of the word. The resulting form of the inflected word is in any case stricter than that of the agglutinating system.
(ii) To the degree that an inflected word form expresses its syntactic function in the sentence it is not forced to construct its sentences by typically determining syntactic role specification by word order, selecting for its language type one of the fixed sequences characterizing the language type: SVO-, SOV-, VSO-(S=Subject, V=Verb, O= Object).

As a consequence the poorly inflected language has a basically fixed and function specified word order whereas the consequently inflected language can use the possibility of free word order for emphasizing communication dynamic discourse features. In some languages these features are given dominant influence in discourse. This is particularly true for Slavic languages and also for German. In these languages there is often a tendency for the sentence's first word in the discourse to take up the dominant theme of a given situation as the primary topic. The rest of the sentence may say something about this topic, a rheme. Since it thereby *focuses new information* forthe subsequent discourse the Prague school under the guidance of Sgall and Hajičová gave it the name *focus*. Their subsequent research concentrated on the textual chaining of sentences in which, typically, a topic of each sentence in the chain depends on the focus of the previous sentences, except, of course the topic of the first sentence, which takes up the topic provided by the initial situation.

Let me exemplify some free order possibilities in German. In each case a literal English translation will illustrate the strange sentence structure that is unacceptable in English. The first examples are from Sgall:

Situation: Previous discourse about a particular city. Here is the first remark:

(1)

Possible continuations:

Diese	*Stadt*	*habe*	*ich*	*in*	*mein*	*Jugend*	*nicht*	*gesehen,*
This	City	have	I	in	my	youth	not	seen

(2)

Gestern	*bin*	*ich*	*dort*	*das*	*erste*	*Mal*	*gewesen*
Yesterday	am	I	there	the	first	time	been

or

Dort	*bin*	*ich*	*gestern*	*das*	*erste*	*Mal*	*gewesen*
There	am	I	yesterday	the	first	time	been

(3)
or

Ich	*bin*	*dort*	*gestern*	*das*	*erste*	*Mal*	*gewesen*
I	am	there	yesterday	the	first	time	been

(4)

As Sgall remarks the first word in each of the three sentences (2), (3), and (4) marks a dominant theme "yesterday", "there" and "I." Yesterday indicates a continuation feature for the discourse whereas "there" and "I" take up word meanings in (1).

A more typical case of chaining is the following discourse:

The initial situation somehow thematizes the raiding of the Jeweller's X thus initiating the following discourse, first in literal translation into English then followed by the German text, in the normal German word order:

"Jeweller's X [topic due to the situation before the utterance of the sentence from previous topic. Note, however, that the German noun-phrase is accusative marked!] raided yesterday incidentally the robber-band Y [nominative marked!]. The famous diamond Z [topic because the hearer might assume Z having been at the jeweller's], that accidentally [a topic commentary] on this day also in the jeweller's was, did they apparently [topic specified as a presumption] not remark. In any case [indefinitely marked, since the previous topic is apparent] let they it in the jeweller's."

The German discourse: "Das Juweliergeschäft X überfielen gestern übrigens die Juwelen Räuber Y. Den berühmten Diamanten Z, der zufälligerweise an diesem Tag auch im Juweliergeschäft war, haben sie offenbar nicht bemerkt. Jedenfalls ließen sie ihn im Juweliergeschäft."

The research of Hajičová several mental steps analyzed. The most important aspect for her was the psychological concept that underpins the dynamics in discourse that is particularly flexible in languages with free word order. The psychological concept is assumed to be realized in a *finite mechanism* as if it were analysed in computational linguistics in. But in this book we are certainly more interested in knowing how the mental operation corresponds to processes in the cortex. Interesting tests measured pronoun references in discourse, in locations in which they referred to the topic. The backbone of such a mechanism analysed is that a hierarchy of degrees of *salience* is involved. As Hajičová and her group specified clearly the degrees of salience are modified during the *flow of discourse* (Sgall 2006, pp. 467–468).

Hajičová's psychological description of the communicative dynamism of discourse – in particular when it involves pronoun reference – is as follows: The speaker chooses a more or less specific (or redundant) denomination in accordance with the hearer's disposition to identify the referent of the expression relatively easily. One of the main conditions of this relative ease is determined by the degree of salience (in its neural activation) if the given item in the *stock knowledge* is of *immediate importance for salience*, and high salience is one of the pre-conditions for an item to be referred to by a contextually bound (thematic) expression. Sgall's studies of whether the morpho-syntactic typology of the languages may have fundamental consequences for sentence construction and the possible use in *communicative discourse dynamic* are very interesting, and correlate with Hajičová's analysis concentrating on salience and selectivity guiding acts of communicative attention in the narrative flow of information. Many descriptions suggest a similarity with saccadic eye movement scanning that I discussed in Chapter 3, section 5. There is clearly a fundamental difference of time. In many cases, we may hope that the studies of the complex conditions of practical usage of language may, together with neurocognitive knowledge of brain architecture, help us to find appropriate measurements of details.

8.3 Introduction to linguistic brain analysis: the innate specialized f-knowledge and its ontogenetic development

8.3.1 Jackendoff's problems

The phenomena discussed in the previous section, the dependency of language construal on general background, on prototype families and on efficiency of communicative flow of narration in different language types, indicate that "grammar in life" means complexity of language structure in the frameworks of usage.

Recalling however our task of modelling, and analysing the architectures and processes of language in the brain reminds us that an effective progress of research requires us to proceed from simpler system analysis to more complex frameworks and theories. A list of basic problems to be solved concerning dynamic biological modelling might help as a guideline. Problems of the organization of higher complexity should be left to a future in which the basic problems have been solved.

There is indeed a very appropriate list. It is the one that Jackendoff (2002) presented in his last chapter on pages 422–423. In my view this list is still up-to-date. I agree with the principles, but remark that some terms contain some dangerous implications that need clarification. I first present the list together with introduction:

> "We can considerably sharpen the questions posed by mentalism. Instead of simply asserting that language must be lodged in the brain *somehow*, and debating whether children must come to language acquisition with some innate capacities specific to language, it is now possible to articulate the issues more clearly. Among the questions that have emerged here as critical are these:
>
> - How are discrete structural units (such as phonological features and syntactic categories) instantiated in the nervous system?
> - How are hierarchical structures, composed of such units, instantiated in the nervous system?
> - How are variables instantiated as elements of such structures in the nervous system?
> - How does the brain instantiate working memory, such that structures can be built online from units and structures stored in long-term memory, and in particular so as to solve the problem of 2 (multiple instances in working memory of single long-term memory type)?
> - How does the brain instantiate learning, such that pieces assembled in working memory can be installed into long-term memory?
> - How does one trial learning differ from slow, many trial learning?
> - How is innate specialized f-knowledge instantiated in the brain, and how does it guide behaviour and learning?
> - How is innate specialized f-knowledge coded on the genome, and how does it guide brain development, including the characteristic localizations of different functions?"

Jackendoff adds that these questions are not peculiar to language. The same questions are posed by the visual system and the formulation of action. Language just happens to be a domain where some of the questions have been more a matter of open debate than elsewhere.

In other sections of his book (on p. 22) he also refers to positive perspectives implied in the idea of f-mentalism that is far more tightly bound up with the nature of the 'hardware' than a standard computer analysis but more difficult. It is not enough to concentrate on neurons and single neuron connections only, as connectionism usually does. F-mentalism must also account for the enormous progress in our neurocognitive knowledge of brain architecture.

The consequence is obvious: It is a big error of standard forms of connectionism to *neglect the frameworks of dynamic structure organizations contained in local and distributed architectures of the nervous system.* The first part of the present book provided introductions to components and functions of empirically studied brain models. Neurocognition is indeed a theoretical system based on *functional brain architecture* that is not only concerned with the operative specificity of single neurons but also with the consequences that follow from local and distributed neural cluster architectures, cortical area functions and systems of myelinated fibre fasciculi over long distances in the nervous system. The development of neurocognitive structure analysis is instructive, since it illustrates the original progress of solving problems.

I think that the problems of relating formal structure to the dynamics of brain architectures and processes should not only be considered from the side of theoretical linguistics but also from the side of functional neurocognition. A historical analysis of the problems and the bracketed proposals for solutions show how psychological and biological analyses interacted and developed increasingly complex models based on functional reflection. The reader will easily see that the biological understanding of common connectionist models still relies on the knowledge of the late 1800s. Here is an instructive list:[2]

- The neuron doctrine established at the end of the 1800s defined the principle that individual neurons are the elementary signalling elements of the nervous system. (Golgi and Ramón y Cajal observed that the nervous tissue was a network in which neural cells are connected. They thus proposed the *neuron doctrine*)
- Can biological and psychological concepts be related on the level of discrete areas of the brain – or are the functions of perception, volition or conceiving organized by the complete aggregate field of the brain? (*Lashley* (1950) demonstrated that there is no function of learning and claimed that "area sub-divisions are in large part anatomically meaningless, and misleading as to the presumptive functional division of the cortex." In his view the functional base of the brain is "*mass action*").
- Is there an intermediate position according to which individual neurons are generally *arranged in functional groups* that *connect to one another*

[2] It relies mainly on E. Kandel's (1995) account of the historical development of brain modelling.

in a precise fashion, as already demonstrated by Wernicke's model for language form processing? (Sherrington, Ramón y Cajal and Wernicke)
- Is there an additional intermediate position according to which discrete local regions in the brain perform elementary operations, whereas elaborate faculties are made possible by serial and parallel interconnections of several brain regions? (During the 1980s the convergence of modern cognitive psychology and the brain sciences concluded that all mental functions are divisible into sub-functions. Though we experience mental processes as instantaneous, unified operations even the simplest cognitive task requires the coordination of several functionally distinct brain areas.)

It is a problem of Jackendoff's book that it contains only meagre reference to brain organization and architecture; in spite of its sub-title. It is clear that his list of problems must be considered in relation to the list of the neurocognitive problems. I hope the first part of my book has given some indications and introduced some influential models of modern neurocognition.

Confronting both lists shows that two terms of the first list are implicitly misleading, namely the terms "*discrete*" and "*working memory*". If the dictionaries tell us the meaning of words we learn that discrete ideas or things are *separate and distinct from each other*. We would tend to understand letters and written word patterns in a text as discrete. Even in a syntactic structure description the labels on the nodes are discrete, and several different structure descriptions are discrete. Thus the linguist may think that Jackendoff's structural units mentioned in the first problem statement refer to the notion discrete in this sense. Recalling Figure 1.2 in this book would rather suggest that in the brain there are no discrete units in any sense. Must we forget to look for discrete units in the brain, as Chomsky recommended under the influence of Lashley's position mentioned above in the second problem statement of neurocognition? I think we must not. But the solution of this dilemma is only possible when we take dynamic processes into account, in particular those that describe discreteness of different patterns by binding synchronization. I will return to some examples below. The same is true when problems of token modelling, of variable modelling etc. are involved. Without understanding synchronizing procedures appropriate correspondences of discreteness are impossible.

This leads us immediately to the notion of *working memory*. Jackendoff seems to suggest a mechanism that *builds structures online from units and structures stored* in long-term memory. Does he really mean that long-term memory and working memory are different areas, one for long-term storage the other being the one in which an operative mechanism can build symbolic structures online. This idea is fundamentally wrong in my view. I agree with

Fuster that the difference involved is again something that can only be described in terms of a network in which distributed pieces of knowledge can become active and synchronized *for a short term*. Considering the complete neurocognitive network of pieces of knowledge (cognits) there are those that are momentarily activated and synchronized as patterns and many others are not in any relevant sense activated. The complete system can be called the long-term memory (of pieces of knowledge); those that are active and synchronized in the same *short time interval* are identical with the working memory. It is the memory that is momentarily working, for instance by synchronization. But it may also be working by interactions that lead to different synchronized patterns for the next temporal interval. Thus there is dynamic structure change due to changes of synchronization patterns and not by some object externae mechanism that manipulates discrete and static patterns. In summary: Jackendoff's first, fourth, and fifth problem cannot be solved if we do not learn to model the interplay of dynamic patterns emerging in short-terms in the brain's networks and forget modelling in terms of computational symbol-processing in special areas for "working memory". But as already mentioned several times, the original Jakobson–Teuber principle hinted already at the solution by referring to the self-operations of clusters of neurons.

I am confident to find the ways in which dynamic modelling can be solved. Some examples in later sections will present some constructive model solutions. They will indicate the direction in which we should proceed. But before these illustrations I think that some discussion of Jackendoff's last point, the genome, will contribute to *introductory understanding*.

8.3.2 The genome problem and the development of the perception–action cortex

Here is again Jackendoff's last point: How is innate specialized f-knowledge coded on the genome and how does it guide brain development, including the characteristic localizations of different functions?

We should first try to answer the question with reference to characteristics of brain architecture and developmental dynamic brain modelling. How do they develop normally in the human brain, in particular in the neocortex? Understanding the process of genetic control is by far too complex for an immediate answer. But we can at least define a focus: Broca's and Wernicke's areas organize core characteristics of language. They clearly belong to the perception–action cycle of automatic organization systems in cortex, as already explained in Chapter 2. Thus primary attention should be focused on the possible conditions that the genome would have to control.

Let us recall Fuster's description: In summary there is clearly a *genetic plan* for the development of the entire observable structure of the neocortex. The plan covers all the macro- and microscopic features of that structure, including neurons and their connective appendices – dendrites, synapses and axons. However at every step of development of the expression of that genetic plan, the structural phenotype of the neocortex is subject to a wide variety of internal and external influences. These influences create the necessary and permissive conditions for the normal development of the neocortex and its neural networks. All events of neocortical ontogeny have their timetable. Very important are the developments of the architectural components, the local clusters or columns about which much has already been said. The various developmental stages are well described by Gage and Johnson (2007). It is however important to imagine how long distances in the brain must develop until adulthood.[3]

After having an approximate idea of the complexity of development let me present some details about the cortical development. In the principled structure there is very little variation between mammalian species in the general layered neuron type differentiation of the cortex, while the total surface area of the cortex can vary by a factor of 100 or more between different species of mammal. Thus, in early weeks of gestation, the embryo undergoes complex processes that form the basis for the central nervous system. It is important to note that prenatal brain development is *not a passive process involving the unfolding of genetic instructions*. Rather from an early stage, interactions between cells are critical, including the transmission of electrical signals between neurons. Waves of firing intrinsic to the developing organism may play an important role in specifying aspects of brain structure long before sensory inputs from the external world have any effect (Gage and Johnson. 2007, pp. 418–419).

The connections to the thalamus and loops centred to the thalamus may play an important role in local neural cluster organization, as I will demonstrate in detailed discussions of local cluster procedures The following remark of Gage and Johnson (2007) about histiogenesis is of particular interest. After the birth and the formation of the basic structure of cortical layer, the next neurons to be generated are the pyramidal neurons of the deep layers V and VI, whose axons project to the sub-cortical targets – probably the thalamic loop. The next neurons to be born are the local inter-neurons in layer IV of the cortex. Finally the pyramidal cells of the upper layers, II and III, are generated. They send axons to other cortical areas.[4]

[3] Recall again Fig. 1.8 in Chapter 1.
[4] More details in Sanes et al. (2006) and also in Hatten and Heintz (1999).

These characteristics of cortical development indicate structures that would have to be constructed on genome bases. Since these structures are rather typical for mammals, it should be very difficult to determine the genetics of innate language form specifics. Concerning the innate facts the more developed frontal cortex, together with its close organizational connection to emotional support of communicative object and intention-oriented sound play from 10 months to 18 or 24 months, may even require a kind of "intentionally serious" communicative play in which the better constitution of the human brain compared with other hominid brains allows the child to speak and speak with meaning.[5] Hominoids do not yet have a "serious" idea of denoting and symbolizing perception–action. Their aim seems to be more directed to reduce symbolizing and expressing emotion and social status in communicative context.

8.4 Instantiating grammars in the nervous system

An appropriate approach to Jackendoff's problems must address the first five problems together. The reason is that short answers to single problems may be misleading as long as the other problems are not clarified. Let me start, however, in the order of the problem sequence.

- Problem 1: How are discrete structural units (such as phonological features and syntactic categories) instantiated in the nervous system?

For an adequate analysis we must account for discrete structural units in a more extended sense including simple strings and complex arrangements as for instance formal trees, variables, interface identity indices and also rules or constraints used to define relations.

What we are looking for is a correspondence of situation-used structural units and complexes of linguistics and dynamic properties and activated states in the nervous system. When this is clear we may consider Fuster's general proposal. Psychological, and also linguistic categories and features are understood, in their disciplines, as symbolic representations of *pieces of knowledge.* They should be instantiated in dynamic (!) neural memory networks. Fuster gives a general name to each cognitive and mental piece of knowledge: The name is *cognit.* In Fuster's clear functional perspective there is the following modelling principle: Each cognit – that is each mental piece of knowledge – should correspond to a *neurocognitive network.* We will see later that this idea

[5] J. Bruner (1983) and (1990). In many respects Tomasello's studies followed Bruner's direction. M. Tomasello (2003).

needs, at least on some levels of the perception–action cycle some more sophisticated description. Jackendoff's problem 1 may thus be reformulated as follows: How are linguistic pieces of knowledge and their usage instantiated in neurocognitive networks? The following conceptual elements are relevant: *Categories, concepts, prototypes and features*, shown also in simpler examples with which the constructions of our exercises could start. Jackendoff mentions a few other descriptive entities in connection with his problem statements: We must cope (1) with rules or *constraint*s that serve the purpose of formally generating compositions *in hierarchical structures,* (2) with *variables* occurring in structures, (3) with *token processing.* More units are presented in the subsequent discussion, for instance (4) the *interface indices* and *interface rules.*

Jackendoff's enumeration mentioned a wide list of terms whose modelling in a neurocogntive framework might create problems. Though I am confident that I can solve Fuster's problems in the framework of the model I have in mind *to begin the problem discussion with a simple exercise.* I'll concentrate on *constituent structure grammars*, the *core systems of formal grammar studies,* and the units that occur in them. A constituent structure grammar is formally presented as follows (Chomsky et al. 1963a).

1. *Letters called symbols*, such as 'N' or conventional letter combinations like 'NP' or 'det' or other varieties. There are two types, (1.a) a *constant*, meaning constant denotation, that denotes a specific abstract category or feature of grammar and (1.b) *typed variables*; a variable can be substituted by a constant when the indefinite denotation of the variable should be transformed into a definite one. (This is similar to understand 'a city' as a typed variable and 'Paris' as a definite value for 'a city')
2. *Strings* as a sequence of symbols, for instance the string of three symbols: NP V NP.
3. *Rules* represented by $j \rightarrow \psi$, where j and ψ are strings or single symbols. Formally this representation should be understood as follows:"$\rightarrow$" represents, a finite two-place, irreflexive and asymmetric relation defined on certain strings of symbols read as "is rewritten as". The rule representation with the arrow is called a *grammatical rule* when the symbols to the left and the right are *grammatical categories* (constant or variable).
4. A rule of the form $A \rightarrow \omega$ is a (context-free) *constituent structure rule.* If a system of rules determining a grammar contains only (context-free) constituent structure rules it is a (context free) *constituent structure grammar*. Remark; There are also context-sensitive grammars; since they are not relevant for our present study their definition is omitted.

This type of formal grammar is an offspring of mathematics, in fact a special case of the class of formal systems that were introduced in the 1930s to define "finitary" mathematical systems.

This formal type of grammar, which Chomsky introduced more specifically for precise descriptions of a core structure for syntax, served since the early 1960s as an exercise frame. At that time it was the modern "trivium" for linguists interested in formal grammar before they could move to study more complex and linguistically more adequate grammar representations. Though used for practising we must acknowledge that this grammar type is not a mere construct. In a sense, basic varieties of languages are of this type. *Basic Varieties* are the kinds of pidgin languages that second language speakers normally develop on their own without any explicit teaching (Klein and Purdue 1997).

The traditional convention of saying that a rule "application" can be read as "is rewritten as" (see number 3. above) was conducive to viewing the rules as like a programme for constructing sentences. In parallel a grammar of this type was called "generative grammar", a term that suggested that grammatical structures were generated. The experts strictly emphasized that these grammars were formal systems in which structures are *formally defined* and not in any way processed. They insisted, that they were after all describing competence or knowledge, not performance. In this way they tried to make students familiar with the abstract formalist perspective.

On the other hand, Chomsky himself had already invented a more transparent representation for coherent rule combinations. It was the representation in terms of *labelled trees*. If one says for instance – not quite neatly – that rule applications generate a sufficiently simple sentence it is possible to represent the constituent relations by branching, as was known from mathematical graph theory and called a tree. Chomsky's early example was: (1) "The man hit the ball", a primitive sentence indeed.

Its syntactic structure is generated by the few syntactic rules presented in Figure 8.1a. Each constituent structure rule is applied by replacement of the symbol on the left side of the arrow when it occurs in a given string of symbols. Obviously the categories NP and VP have a double role in syntax; they appear left and right, though in different rules.

The formal application procedure starts with placing the symbol S (denoting the category Sentence) in the "working space". This provides the condition for applying the rule with S on the left of the arrow, thus generating the sequence NP VP, with which other rules can proceed. The steps that result from applying one rule after the other lead to the sequence of replacements presented in Figure 8.1b.

The reader will easily verify the similarity of this arrangement with the configuration of a labelled tree branching that Chomsky invented as a practical

a

S	→	NP VP
NP	→	Det N
VP	→	V NP
V	→	hit
Det	→	the
N	→	man
N	→	ball

Figure 8.1a. A simple set of syntactic rules.

b

Start condition:	S
Rule (1); S replacement:	NP - VP
Rule (3); NP replacement:	Det – N – VP
Rule (2); VP replacement:	Det - N - V - NP
Rule s(3); NP replacement:	Det - N - V - Det - N

Figure 8.1b. Rule applications as symbol replacements.

c

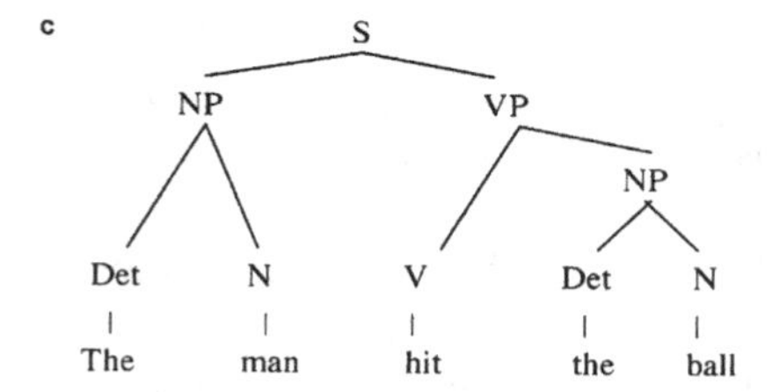

Figure 8.1c. The syntactic tree assigned to *The man hit the ball.*

form of representation for linguists. Being equivalent to Figure 8.1b, Figure 8.1c serves the purpose. In both representations the character of hierarchical arrangement of categories is transparent, as well as the part of speech sequence at the bottom of the tree.

These were different schemata for representing context-free constituent structure grammars based on rule relations formally defining a recursive hierarchy. They connect and sequentially combine single symbols and strings of symbols. The linguistic specification of syntactic structure is clear and strictly precise in the framework of a formalist theory.

Having selected them as our first study objects we are now confronted with the question of how rules of formalist grammar and defined structures could be appropriately incorporated into a *model of neural performance*. In computational linguistics there are many models for programming computers. But from our point of view all of them are in principle misleading. It is their basic architecture and organization that radically contrasts with the brain's system. My aim is to specify a neurocognitive model that relies as much as possible on empirical knowledge about the cortex and sub-cortical and relevant annex components. Many of relevant characteristics have been discussed in Chapters 1 to 4. The following section will take up the challenge of translating formal grammar into empirically plausible neurocognitive network models.

8.5 Translating trees, rules and rule complexes into neuronal modules

My first idea was the attempt to solve the problem by generalizing the Jakobson–Teuber principle, as presented in Chapter 1, section 1.3:

> The distinctive features of phonology, the linguistic unit denotations, and the linguistic categories on other levels are no longer understood as symbolic elements of universal notional schemata for classifying phonemes, or structuring words and sentences in all their diversity across languages. Instead, linguistic features, units and categories are to be considered as "real" in the sense that each is represented by universal interacting neural cluster mechanisms, contributing to organizing dynamic patterns in the brain's cortex, understood as generating and controlling form, meaning, and context interdependency of all elements of speech perception and production.

I think that the basic translation principle is still correct: Assign a "neural cluster mechanism", that is a neural module unit, to each of the linguistic features, units and categories that are functionally specified in a formal grammar – here a constituent structure syntax. But it will be necessary to add more specificity as soon as the analysis is confronted with the problems that Jackendoff correctly enumerated. I will handle these challenges later, after having once more repeated the *general facts to be known about the modules*. Each neural module is an ensemble of some pyramidal neurons having specific connections to other components in the brain. They constitute the possible activity relation to other modules in different components of the brain; they also keep the interaction patterns representing the momentary presence of a complex structure of different pieces of knowledge, that is cognits. This synchronization is partly determined by interactive fit and partly by organization support of units of the limbic system, especially the thalamus and some units interacting with the thalamus. The reader should again study what was said in the first chapters, in particular Chapter 1. Let me recall just a few characteristics. The modules are sometimes identified with so-called columns that contain closely related neurons, such as for instance visual cells that correspond to different orientations and also units characterizing basic sound distinctions underlying phonological features. But more generally columns may also be clustered into hyper-columns, which may be part of an even larger cluster. As already mentioned elsewhere the range of cooperating clusters is usually larger the more they are distant from the phyletic level of the perception–action hierarchy. They are in any case larger in the multi-modal or trans-modal level and even larger in the pre-frontal cortex

The core elements and the primary connections are in any case established by the pyramidal neurons, the "parent" neurons. They are always surrounded by inter-neurons. When the range of parent neurons is larger the system of inter-neurons is also more differentiated in its task of supporting the general connectivity in activations. The complete cooperation of pyramidal neurons and inter-neurons is not yet understood in the details, but it is clear that their cooperation has important power in generating new connections and in synchronizing established ones. There is a complex topography in the perception–action system, like in geography. In this metaphor the brain is like the world, certainly also in its dynamic possibilities of interaction. New streets are built and old ones move or are rebuilt. This is also true in the adult brain. Neurons continue to grow, migrate, connect, disconnect and die, even in the healthy mature brain. The brain is never frozen into a rock-like state.

8.6 Meeting the challenges of bridging the gap

I believe that the much more complex brain architecture on the micro- and macro-level provides dynamic elements, which allow us to construct neurocognitive working models. Neural networks should represent specific interactive capacity which, when activated in contingent contexts, generate short-term activity patterns in sub-sections, based on their binding competence. The merely formal structure-generating systems, which we want to translate into specific working model patterns, are given by linguistic knowledge as presented above. Its symbolic frameworks must somehow be brought in relation to neurocognitive model frameworks. These models usually envisage realizations whose organization relies on modules, which are local clusters of neuron circuits. The most important neurons in the circuit are pyramidal cells as centres. Their axons have cortico-cortical connections to other modules and also connections to sub-cortical organization centres, such as the thalamus. Their operation is differentiated by interaction with surrounding inhibitory or excitatory inter-neurons with functionally appropriate cooperation.

The first ideas for translation assume that (a) certain elementary *modules*, that is local network clusters, represent elementary linguistic categories[6] and

[6] Recall the Jakobson–Teuber principle.

that (b) *momentarily synchronized interactive modules* represent utterance underlying *structures,* phonological or syntactic. Note that this assignment does not imply that the corresponding complex ensembles are clearly biologically marked as separated networks. There is some probability that this is true for feature distinctions on phyletic levels in which external signals from the perception or action organs contact cortical input and output. In the range of linguistic form they organize phonological feature groups. On the other so-called association levels complex assemblies may often overlap. Certainly, possible momentary active structure representations – corresponding to linguistic structure knowledge – must not be generated in the momentary activation by separate brain sections. It is sufficient that the *momentarily appropriate* sub-network becomes active while other overlapping patterns must wait for their selective context activity. The single modules brought into an appropriate state must compute their synchronicity by the cooperation of their "neural computational power" in which hundreds of local cluster neurons, pyramidal cells together with inter-neurons, check the incoming and outgoing contacts with other clusters. Given this complexity our functional analysis may well suppose that the dynamic power of interactive module processing is sufficient to functionally disentangle the superficially perceived muddle, as it was illustrated in Figure 1.2 in Chapter 1.

Recognizing modules as the proper units for representing features and categories *contradicts the common assumption of connectionism.* According to our working model *categories are not represented by single neurons*, not even by only one singular pyramidal neuron supported by some inter-neurons. We need stronger clusters but must acknowledge that in the present research situation the precise details of the modules' organization are not yet known. The general possibility is given and encourages further studies in model constructions or in much more difficult measurements of types of modules. Certain modelling designs have been proposed in various directions. The studies of model structure by Jeff Hawkins (2005, Chapter 6) are very instructive.

After having indicated how features and categories, generative rules and constraints could be transformed into dynamic network possibilities and momentary network synchronizations that correspond to temporal intervals of use we are now ready to confront the specific problems Jackendoff enumerated. The *first step* must be to eliminate the misleading conceptualization of the *working memory.*

We should not be guided by ideas of artificial intelligence or computer programming. I emphasized it already at the end of the formal definition of constituent structure syntax. I insist that the brain has *no central processor unit*, no *constructive separation* of central processor and spaces for long-term or short-term symbol arrangements and data structures – as is usual in symbol-processing computation systems, *no transport of symbolic units or complexes*, and no *symbol-combinations, separations or rearrangements* in these spaces. The essential error of the traditional computational approaches derives from a wrong concept of "dynamic" in which passive non-dynamic objects – symbols and symbol patterns – are separated from an operative component that operates on the symbol patterns are determined by symbol-operating rules.

This idea is unfortunately also applied in the *classical concept of working memory*. Its organizational description is as follows: There is the assumption of an operative central processing unit (CPU) that accesses a separate *long-term memory* in which symbol strings and data structures are stored. The CPU selects some of them, places them into another space, the *working memory*, and operates on them in the working memory by combining, separating, moving etc., until appropriate and identifiable structure patterns for later access are generated and transferred back to the long-term space.

It is true that even some neurocognitivists use these ideas and terminologies. Based on his detailed knowledge of memory in the cortex Fuster 1997 has criticized these models. I completely agree with his view. In it the specific connectivity of neurocognitive networks has two simultaneous dynamic functions, first keeping the possibility of generating specific activity patterns, waiting until contextual information causes activation. Next the causal interacting activation contributes to neurocognitive binding activity, synchronizing the activity patterns where they are for the short term. As Fuster says there is no particular brain *location of short-term memory* for the task of processing. All relevant sections of the complete cortical network of the perception–action system *can function at their place* as an active *momentary memory for the short term*. The next sections will explain how this is organized in the particular case in which the neurocognitive function corresponds in its activity structure to the rule-determined structure of constituent structure syntax.

Having explained the possibilities of constructing working models I will now turn to discuss the concrete challenges that are often presented by critical analysts concentrating mainly on *fundamental problems* that *Jackendoff thankfully enumerated and explained*. In principle I'm confident that Jackendoff's problems can be solved. His discussions show that his problems consider the *typical cases of connectionism*.

Though being optimistic to solve the problems by my models it seems to be more appropriate to proceed carefully in many steps from simple to more complex constructions. In the present chapter I will *concentrate on* translations of formation rules and more specifically *constituent structure systems*. The explanations promise to show how certain problems that are not specific for formation rules are implicitly involved. This concerns Jackendoff discussions of *typed variables* and of *learning problems*. As already mentioned his scepticism is mostly substantiated by considerations of spreading activation systems and *standard connectionist models*. In a summary statement he writes for instance that it could be shown that for principled reasons the models of Rumelhart, McClelland and Elman cannot encode variables of the sort, necessary for two-place relations such as "X is identical with Y", "X rhymes with Y" and "X is the regular past tense of Y". This principled failure is fatal to unadorned spreading activation models of language, for all combinatorial rules – language formation rules, derivation rules, and constraints – require typed variables (Jackendoff 2002, p. 65).

This statement is important since it suggests that an appropriate neurocognitive organization of formation rule systems may indicate how the problems of typed variables and typical learning differences are implicitly solved in formation rule systems. Therefore I will not continue to discuss the other problems. Concentrating on the translation of formation rules is more important for understanding their translation result in interactively active neurocognitive modelling. The details of translation will also give us the chance of explaining synchronization problems in neurocognitive constituent structure network systems.

At the same time I will solve Jackendoff's problem 2, asking how hierarchical structures are instantiated in the nervous system, and problem 4 that asks *how tokens are organized.* Explaining synchronization and the processing of tokens is the core of the following section. The solution of both problems demonstrates well the importance of the internal processing of modules.

8.7 Organizing binding and tokens in neurocognitive constituent structure networks

Concentrating on Constituent Structure Grammars as relatively simple formalistic systems it is still helpful to ease our understanding of other varieties of structure representations. Indeed constituent structures cannot only be represented by a set of rules. A simple alternative is Joshi's approach (1987). He uses so-called treelets, schemata that help one to think intuitively about structure relations. In our case the three treelets of Figure 8.2 may illustrate their form and use. They correspond to the three rules that also generate the constituent structure grammar of Figure 8.1b above belonging to the sentence *The boy hit the ball.*

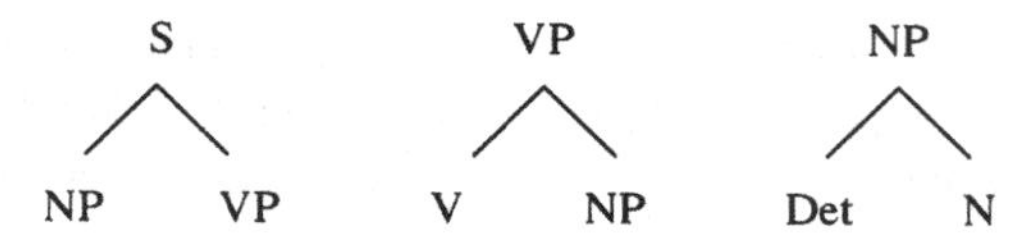

Figure 8.2. Three tree lets whose clipping leads to Figure 8.3.

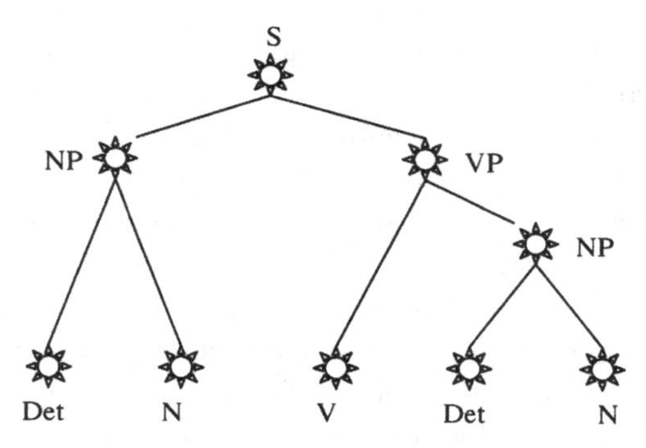

Figure 8.3. Clipped tree transformed in module arrangement.

The constituent structure of Figure 8.3 is generated, when clipping the NP twice, first to the S-treelet, and second at the NP symbol contained in the VP-treelet. The resulting tree topology is in principle the same as the labelled tree of Chomsky's original sentence "The boy hit the ball!" in Figure 8.1c.

In Figure 8.3 I used a *new type of node representation* signalizing the first step of my translation into a neurocognitive network. Representing the nodes as stars should indicate that they *would represent*, in the ultimate constructions, *elementary neuronal cluster modules* having *potential interactivity relations that in complex networks can interact in many appropriate directions*, bottom-up, top-down, collaterals, and possibly other alternatives.

Unfortunately, reflecting our generalized Jakobson–Teuber principle shows that the simple tree in Figure 8.3 is wrong. The principle definitely required that each symbol denoting a category type be translated into exactly one local cluster or module. The principle excludes the repetition or recurrence of categories; it does not allow several tokens of the same symbol! What should we do? We take the S-treelet and clip the NP-treelet to it – indicating the sentence subject at this location. And then we clip an NP treelet to the VP thus representing the "object" NP. But stop. This conflicts with the Jakobson–Teuber principle. There are now two NP tokens instead of a single NP for the NP category only.

The token-avoiding solution is as follows: A category node must be assigned a module and the internal cluster circuit of this module must be able

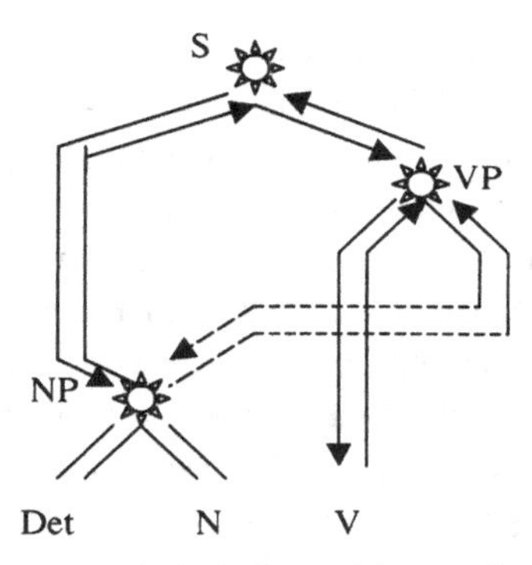

Figure 8.4. Replacing the use of tokens by module organization.

to distinguish internally, for a short time, the moment in which it is active as an "object" NP from the moment at which it is active as the "subject" NP. If we have such a module we no longer are forced to use tokens. It is possible to transform the tree of Figure 8.3 into a configuration in which the NP node has convergent connections as in Figure 8.4 such that the node receives alternative signals from above, that is from the S or from the VP node.

The processing consequence is clear. Each node is not merely a component for signals of a single higher node. As in the case of our NP node the new activation coming from S gives the NP the syntactic function of the subject part whereas the activation from VP marks the function of a grammatical object. We now suppose that the NP node represents an internal circuitry of a neural module, usually a network of several hundred neurons. In this case it is a powerful memory. It organises *local short-term memory* for input alternatives and thus instantiates a memory specific dynamic neural module.

This capacity explains also that module circuitry can in principle serve more complex tasks in managing different connectivity operations, not only combinational activity in the perception–action hierarchy. Based on these modules it will be possible to solve the token problem to which we will return. We will moreover envisage a spectrum of different kinds of activity. For instance, the modules must be able to react appropriately to *neural collateral information* of the multi-modal and poly-modal connections, and to the many influences that were informally discussed in the first part of this chapter, *mind-space* and *background information*, *selections of prototypes* and the role of *salience based selections* in constructive narratives.

These tasks cannot be solved by single modules but by functionally efficient cooperation of many modules. But an appropriate understanding of module internal circuitry is not yet possible.

In this chapter I select one particular challenge, namely the possible organization of tokens in complex forms of active and synchronized structure representations, activated as sub-sections of networks. Our challenge is to describe token difference of subject and object NP, as mentioned above. Our knowledge from the hierarchies in the perception–action system for language that we presented in Figure 2.3 of Chapter 2 indicated that there are reciprocal connections, bottom-up as well as top-down, as already represented in Figure 8.4. If activations over these reciprocal connections are first considered in the bottom-up direction, the module should have some internal decision competence for appropriately sending the signals either to S or to VP.

We must now start with the discussion of processing details. Consider inputs from top and follow the discussion of some details. I shall not explain the solution in all of its details but the reader should at least get an idea about the dynamic process patterns that result from our translation into hierarchy distributed modules. Here are again the essential ideas to be applied to the network with overlapping NP; these ideas rely on connectivity determined signal flow. First, the static configuration of a tree is translated into a dynamic process that is generated and controlled by the internal circuitry of the modules. Second, as already mentioned, all lines are reciprocal and thus doubled. Let me now sketch the bottom-up situation. The reader is invited to follow the description, by referring to Figure 8.4.

In the beginning of a sentence an initial utterance sound is heard and causes an expectation signal flowing along the top-down connection from the S-module to the NP-module, and from there to the Det module. Here now is an *important point*: In the NP-module, the internal circuitry of the module (that is network cluster for NP) *stores for a short period* the fact that the input signal came from S. The Det module registers by neuronal analysis that the initial sound pattern "the" is indeed a Det. This causes a signal sent along the bottom-up connectivity line (recall the reciprocal pair of connections) to the NP-module. Since this module corresponds to the NP rule, its circuitry registers the completion of Det and concludes that an initiating expectation signal should be sent to the N-module. This registers that the sound "boy" indeed satisfies the expectation and consequently sends a completion signal back to the NP-module. Here now is the *next important point* depending on cluster internal short-term memory; registering the NP completion, the registering signal now "asks about the intermediately stored input state" from a short-term memory component of the clusters circuit's "storage sub-network". The stored short-term activity indicates that the original input came from S. The combination of this storage signal with the NP completion signal now generates a signal forwarded to S. The module S has an internal circuitry corresponding to the conditions of the rule or treelet. It determines that after completed NP, VP should be sent an initiating signal from the S-module. It causes VP to signal grammatical expectation to the V-module. V registers that "hit" is indeed a V, a fact that causes it to send the completion signal to VP. Corresponding to the rule, VP circuitry now sends an expectation signal to...? Well to NP! There is now again an *important point*: NP gets for a second time an expectation signal, but this time from VP. NP's circuitry memory for a short time now stores having received a signal from the VP input connection, NP now sends an expectation signal again to Det, which registers the presence of a Det, sends the completion signal to N, "receiving the sound signal "the ball", which after reception causes, corresponding to the rule, signal to NP. As expected there is now the next important point: The NP short-term store, having registered an input from VP, now sends a completion signal to VP that in turn sends a completion signal to S. Since the connections in the modules are short one can assume that any act mentioned takes about 25 milliseconds, the sum about 400 milliseconds, less than the normal speech length for the sentence.

Careful attention focusing on this structure "scanning", in particular the short-term storage in local module internal neuronal operations should have explained that the system *operated two NP tokens* by short-term memorizing *distinct input directions* from the neighbourhood. The first short-term store marked the ongoing registration of the first NP token between the moment of the signal coming from S and the moment of the N completion signalizing the completion of NP. This short-term storage interval marks the *"subject"-NP token*, the one that was structurally activated from S. At the next "important point" the short-term storage marks input from and output to the VP process, thus *marking the NP-token that is the "object"*. This demonstrates that module internal networks for short-term storage of network directions are activated by their input and de-activated by their output. *The token problem, that is the problem of 2, is thus solved.*

Let us now discuss further use that results from this processing. It may be that the sounds of an utterance heard are not very distinct, even having perhaps an unclear meaning. In this case the "scanning" of the network could be repeated one or several times; activating in this way inhibitory signals. Indistinctiveness of sound features might by clarified by repetition. Note that in this way of "repetitive structure scanning" the periodic attractor *generates itself the binding* of the translated constituent structure grammar. *This solves the first of Jackendoff's challenges*. I emphasize again the essential features of Fuster's extensive discussion of the binding process (Fuster 2003, pp. 99–100). There is a clear measurement of High Frequency activity (HF) in the cortex. This is certain and one assumes that the measurements somehow indicate binding processes, but there is no agreement how the generation of HF-activity is generated. There are two alternatives none of them being yet clearly established. The first assumes that there is an inherent tendency of neurons to oscillate in certain brain structures, notably the thalamus, and that cortical HF can be a manifestation of reverberation and cortico-thalamic loops. Others take the position that HF activity in the cortex is generated in the cortico-cortical circuits and has encoding properties. Fuster maintains that the internal network activities entail the synchronized firing of its neurons. Depending on the networks cortical neurons tend to drift in and out of certain frequencies and temporal patterns. Fuster views these frequencies and patterns as fixed-point attractors, that is states that are defined by the functional architecture of the net and to which its cells gravitate. It is obvious that what I said before is in better correspondence to Fuster's position than that of the others.

This was just a first sketch of how the static constituent structure tree is translated into a neuronal network model that makes use of typical properties known from neuronal networks: no token repeated symbol tokens in a string will be found in the brain. Instead, the local cluster organizes reciprocal connectivity and convergence. Discussing the fact that more complex rule system translations will lead to divergence, multiple co-laterality and choice points requires further development of modelling that transcends our initial neurocognitive model exercise translating a constituent structure syntax.

8.8 The power of the nodes in widely distributed types of connection areas

Solving the organization of tokens by short-term memory in a category representing a cluster of neurons suggests that the local clusters of neurons containing many pyramid neurons – parented units and probably many hundred interneurons in their local environment – can organize many specifications of contact, for instance with other clusters representing relevant and meaningful context data. The specification of contact registration gives the category cluster a large number of potential differences of particular directions that may be activated contingently by short-term activity fixation. This is for instance the case when the cluster represents an ambiguous word. The context data give it an appropriate direction for meaning selection. In another respect we may consider how a cluster could *organize a variable*, which the cluster should represent. We should distinguish two types, the consciously selected variable value and the selection of the variable value that is automatic in the normal context. In the first case the cluster should have normal contacts with thought organizer areas that would send selective signal distinction to the variable-representing network. There may be two parallel signals; one indicating that a variable value has been *selected in one of the thought areas of the pre-frontal cortex* whereas the other signal selects the value signal. In the automatic network the process is similar except that the appropriate value signal has been automatically selected in the systematic network. What was just explained emphasizes the possibility of representing connections and relationships that are emotionally and intellectually important, thus basically relevant and meaningful for the mind/brain.[7]

Let us briefly consider Jackendoff's question how the brain decides whether two words rhyme. Let us take Jackendoff's Yiddish word pair "Oedipus – Shmedipus". The brain organizes the phonotactic structure of the words it hears, one after the other. It may even be that the brain repeats several times the word sequences it hears. Obviously in each case a distributed phonotactic category pattern is temporally synchronized by a number of phonotactic categories. Two synchronized patterns become short-term active one after the other. If the repeated activation of the two words is direct, the following happens: The onset parts register quickly from one into the other whereas the rhyme parts remain the same in phonetic repetition. The brain signals a direct rhythm! Thus the experience of a rhyme is automatic and absolutely simple. The example shows that synchronized prosody pattern comparison may have much influence in feeling experience. This is a good exemplification of Fuster's statement: "Reverberation through recurrent neuronal circuits is a likely mechanism of working memory and therefore of temporal integration. Consequently working memory appears to be a *mechanism of temporal integration* based on the *recurrent activation of cell assemblies* in cortical long-term memory networks."

I will not continue to discuss further examples. The essential point is that the brain experiences relations between synchronized activity patterns that are either simultaneous or temporally ordered activity cluster distributions. In many cases there are rhythms correlations that generate another or a similar synchronized pattern. Working memory is the process that works in a short time activating comparative selection procedures, perhaps similar to a saccadic scanning of vision field, and generates an information conclusion as a new activity pattern. It is not excluded that the reaction to comparison leads to activating a complex activity pattern in the network that was not active before. The result could sometimes be understood as a synchronized pattern combination of two

[7] Recall the discussion of the meanings of *meaningful* and *relevant* in section 7.4 "The fourth stage of pushing the world into the mind/brain/body" in connection with Figure 7.3 on the level "Integration of knowledge bases and knowledge dynamic".

previously synchronized patterns. But this was not the result of moving two static objects into contact. It was instead an internally causal relation among momentary activity parts of the brain. It should be clear that the local clusters of categories or word units have sufficient processing power to react appropriately on given influences of backgrounds, prototypes and topic-focus support that the first sections of this chapter presented as elements supporting fluent understanding of common sense complexity.

8.9 This–There versus Now and Then; combined with some concluding remarks

The present chapter has concentrated on the dynamic of language organization in the brain. Due to the challenges of theoretical linguistics the more extended sections concentrated on the hierarchical organization of a simple form of syntax. The chapters of the first part had already emphasized that this form-organization of language is in close relation to meaning organization, either formal meaning relations which are also organized in the brain sections of form organizations or concrete meanings referring to aspects of perception and action or feeling. The organizations of the latter are widely distributed in the brain. Already the discussions of grammar in the life in Chapter 6 have indicated that further aspects of efficiency, flexibility, objectivity and subjectivity in concrete speech act require appropriate organization. Section 8.1 extended these perspectives. The introduction of the roles of background, prototypes and communicative dynamics of discourse differently organized by free or rigid word order explained various supporting types of speech understanding.

All of this explained the essential importance of dynamic, as it is active in uses of speech acts or inner causality of organization modules. But what is still lacking is the understanding of principled characteristics. In some sense the translation of symbol configuration grammar into interactive activity module networks indicated that concentrating on the organization per se may change the theoretical value from abstract entity knowledge to the concrete world of what happens or what is felt as the organism's inner drive.

If this aspect of dynamic is seen as fundamental it is no longer understood in the sense of present physics but rather in the sense of Aristotle and Leibniz. Leibniz (1714) stated that every created being is subject to continuous change physically and mentally. Even the understanding of macro- or mini-components is revised. This fundamental perspective suggests that the organism and its components generate the changes contingently. At any moment they have the possibility, that is the δυναμισ, for change into an active that is energetic state, but it may well wait for the appropriate moment. When the conditions are given, they change from "waiting" for a state of being energetically active for a momentary appropriate interval. The animals' bodies and brains are typical examples and psychic and mental experiences feel it in their experience of life and self. In this view the basic ideas are concentrated on *what happens now* and what *happens then* and which is the character of the continuous flow of experiences and their principles.[8] Our examples considering language in mind and brain tried to encourage research for developing theory and modelling in this sense of dynamic.

As an interjection I may remark that our perspective is confronted with the controversial stance, the standard view that the world and the body consist of distinct and identifiable configured objects properly understood as a configuration of static pieces of knowledge. The latter view hints at artificial intelligence (AI) in which the cognitive spaces of symbol objects are the knowledge world. In the first case agents operate on objects in the world, in the second case central processors (CPUs) operate on symbol objects in their "world". The AI view suggests a similar view for brain understanding. Its knowledge pieces are the symbol objects in storage spaces instantiating long-term – or short-term memory. In all AI cases we have passive data in passive data spaces on the one hand and central operators on the other. The former are dynamic, the latter strictly passive. Meta-mathematical theories of formalized symbol processing of the 1930s provided the basic idea for computer theories and constructions at mid century. These ideas were overwhelming and influenced modelling and even provided interpretation frames for experimental cognitive science.

Quasi-biological world views as they were presented by Aristotle, Leibniz, Spinoza, and principled dynamist designers of modern neurocognitive models like Fuster and Damasio do not see any passive units in biological animals and not any central processor unit. Self-dynamic units that are more or less energetically active in their interactions constitute the animals' bodies, including their nerve systems. And their counterparts, the inner drive determined units that in human individuals attain the status of minds are equally dynamic and are energetic, if the situations are appropriate.

That should be enough for the philosophical background. Let us return to language and neuroscience. Focusing on the former, we are looking for a new interpretation. Language should be described as a dynamic competence, mentally activated in the intentional energy of speech acts and also historically

[8] Damasio's schematic "sketch" representation of core consciousness is instructive – (1999), p. 178 and philosophically further substantiated in Damasio (2003).

changing in its social *energeia*, as Humboldt said. These dynamic views should in principle be better substantiated by the analysis of *language in the brain* than *language in symbolic formalisms.*

Which are the essential elements of neurocognitive understanding? We should definitely not forget the fundamental integration of our brain organizations. It contains the automatically self-organizing sub-systems, the perception action organization system and the body internal autonomous and somatic nerve-systems and their integration organization in the pre-frontal cortex, and the nervous system centres of hypothalamus and thalamus.

Though language form organization is only a section of the perception–action system it is closely connected with practically all other sub-systems that contribute in many ways to our understanding and self-understanding, conceptually focused in selective ways.

This is the most important point: Given the complexity of our bodies and minds and given our selectivity in consciously focusing and literally or ritually fixing what we consider as basic, we normally do not acknowledge that what we know consciously is necessarily only a skeletal system of what seems to exist here and there but is supported by a much more detailed infinity of elements constituting the flow of now and then.

References

Arbib, M.A. (2002). Grounding the mirror system hypothesis for the evolution of the language ready brain. *In* Cangelosi, A., and Parisi, D., editors. *Simulating the Evolution of Language*. Springer, Berlin.

Arbib, M.A. (2003). Language evolution: the mirror system hypothesis. *In* Arbib, M.A., editor. *Handbook of Brain Theory and Neural Network*, 2nd edition. MIT Press, Cambridge MA.

Arbib, M.A., Erdi, P., and Szentagothai, J. (1998). *Neural Organization: Structure, Function, and Dynamics*. MIT Press, Cambridge MA.

Arnheim, R. (1969). *Visual Thinking*. University of California Press, Berkeley CA.

Arnheim, R. (1971). *Entropy and Art*. University of California Press, Berkeley: CA.

Baars, B.J. (2007a). A framework. *In* Baars, B.J., and Gage, N.M., editors. *Cognition, Brain, and Consciousness – Introduction to Cognitive Neuroscience*. Academic Press, London, pp. 31–58.

Baars, B.J. (2007b). Language. *In* Baars, B.J., and Gage, N.M., editors. *Cognition, Brain, and Consciousness – Introduction to Cognitive Neuroscience*. Academic Press, London, pp. 317–341.

Bar-Hillel, Y. (1950). On syntactical categories. *Journal of Symbolic Logic* **15**:1–16.

Bar-Hillel, Y. (1954). Logical syntax and semantics. *Language* **30**:230–237.

Baddeley, A.D. (2002). Fractionating the central executive. *In* Stuss, D.T., and Knight, R.T., editors. *Principles of Frontal Lobe Function*. Oxford University Press, Oxford/ New York, pp. 246–260.

Ben Shalom, D., and Poeppel, D. (2008). Functional anatomic models of language: assembling the pieces. *Neuroscientist* **14**:119–127.

Blakemore, S-J., and Frith, U. (2005). *The Learning Brain*. Blackwell, Malden MA.

Brauer, J., Neumann, J., and Friederici, A. (2008). Temporal dynamics of perisylvian activation during language processing in children and adults. *Neuroimage* **41**:1484–1492.

Bruner, J. (1983). *Child's Talk – Learning to Use Language*. Oxford University Press, Oxford UK.

Bruner, J. (1990). *Acts of Meaning*. Harvard University Press, Cambridge MA.

Cannon, W.B. (1939). *The Wisdom of the Body*. Norton, New York.

Caplan, D. (1999). Language and communication. *In* Zigmond, M.J., Bloom, F.E., Landis, S.C., Roberts, J.L., and Squire, L.R., editors. *Fundamental Neuroscience*. Academic Press, San Diego, pp. 1499–1502.

Carnap, R. (1937). *The Logical Syntax of Language*. Routledge and Kegan Paul, London.

Carnap, R. (2003). *The Logical Structure of the World*. Open Court, Chicago, Ill. (1928 German ed.).

Chafe, W.L. (1973). Language and memory. *Language* **49**:261–281.

Chafe, W.L. (1974). Language and consciousness. *Language* **50**:111–133.

Changeux, J.P., and Danchin, A. (1976). Selective stabilisation of the developing synapses as a mechanism for the specification of neural networks. *Nature* **264**: 705–712.

Chomsky, N. (1957). *Syntactic Structures*. Mouton, The Hague.

Chomsky, N. (1963a). Introduction to formal analysis of natural languages. *In* Luce, R.D., Bush, R.R., and Galanter, E., editors. *Handbook of Mathematical Psychology*. John Wiley, New York, pp. 269–321.

Chomsky, N. (1963b). Formal properties of grammars. *In* Luce, R.D., Bush, R.R., and Galanter, E., editors. *Handbook of Mathematical Psychology*. John Wiley, New York, pp. 326–418.

Chomsky, N. (1965). *Aspects of the Theory of Syntax*. MIT Press, Cambridge MA.

Chomsky, N. (2000). Linguistics and brain science. *In* Marantz, A., Mijashita, Y., and O'Neil, W., editors. *Image, Language, Brain*. MIT Press, Cambridge MA, pp. 13–14.

Damasio, A. (1994). *Descartes' Error – Emotion, Reason, and the Human Brain*. Grosset/Putnam, New York.

Damasio, A. (1999). *The Feeling of What Happens – Body and Emotion in the Making of Consciousness*. W. Heinemann, London.

Damasio, A. (2003). *Looking for Spinoza – Joy, Sorrow, and the Feeling Brain*. Vintage, London.

Davidson, D. (2006). *The Essential Davidson*. Oxford University Press, Oxford.

Deacon, T.W. (1998). *The Symbolic Species: The Co-evolution of Language and the Brain*. Norton, New York.

Dehaene S. (2009) Signatures of consciousness: a talk by Stanislav Dehaene. The Third Culture*: Edge* in Paris. Available at: http://www.edge.org/documents/archive/edge 306.html. Accessed 1 December 2009.

Dehaene, S., Kerszberg, M., and Changeux, J.-P. (1998). A neuronal model of a global workspace in effortful cognitive tasks. *Proceedings of the National Academy of Sciences of the United States of America* **95**:4529–14534.

Dehaene-Lambertz, G., Hertz-Pannier, L., Dubois, J., and Dehaene, S. (2008). How does early brain organization promote language acquisition in humans. *European Review* **16**:405.

Dennet, D.C. (1995). *Darwin's Dangerous Idea*. Touchstone, NewYork.

Einstein, A. (1961). *Relativity – The Special and the General Theory*. Three Rivers Press, New York.

Erikson, E.H. (1950). *Childhood and Society*. Norton and Co, New York.

Erikson, E.H. (1977). *Toys and Reasons – Stages in the Ritualization of Experience*. Norton and Co, New York.

Ferrari, P.F., Gallese, V., Rizzolatti, G., and Fogassi, L. (2003). Mirror neurons responding to the observation of ingestive and communicative mouth actions in the monkey's ventral premotor cortex. *European Journal of Neuroscience* **17**:1703–1714.

Flechsig, P. (1920). *Anatomie des menschlichen Gehirns und Rückenmarks*. G. Thieme, Leipzig.

Frege, G. (1879). *Begriffsschrift*. Halle, Germany.

Friederici, A. (2002). Towards a neural basis of auditory sentence processing. *Trends in Cognitive Science* **6**:78–84.

Fuster, J.M. (1997). *The Prefrontal Cortex: Anatomy, Physiology, and Neurophsychology of the Frontal Lobe*. Lippincott-Raven, Philadelphia.

Fuster, J.M. (2002). Physiology of executive functions. *In* Stuss, D.T., and Knight, M.D., editors. *Principles of Frontal Lobe Function*. Oxford University Press, Oxford UK, pp. 96–108.

Fuster, J.M. (2003). *Cortex and Mind – Unifying Cognition*. Oxford University Press, Oxford UK.

Fuster, J.M. (2004). Upper processing stages of the perception action cycle. *Trends in Cognitive Science* **8**:143–145.

Gage, N.M., and Johnson, M.H. (2007). Development. *In* Baars, B.J., and Gage, N.M., editors. *Cognition, Brain and Consciousness*. Academic Press, Burlington MA, pp. 418–419.

Georgiades, Thr. (1982). *Music and Language*. Cambridge University Press, Cambridge UK.

Givón, T. (1995). *Functionalism and Grammar*. J. Benjamins, Amsterdam/Philadelphia.

Grünbaum, A. (1969). The meaning of time. *In* Rescher, N., editor. *Essay's in Honor of Carl G. Hempel*. Reidel, Dordrecht Holland, pp. 147–177.

Hatten, M.E. and Heintz, N. (1999). Neurogenesis and migration. *In* Zigmond, M.J., Bloom, F.E., Landis, S.C., Roberts, J.L., and Squire, L.R., editors. *Fundamental Neuroscience*. Academic Press, San Diego CA, pp. 451–479.

Hawkins, J. (2005). *On Intelligence*. Owl Books, New York.

Hebb, D.O. (1949). *The Organization of Behaviour – A Neuropsychological Theory*. J. Wiley, New York.

Hickok, G., and Poeppel, D. (2007). The cortical organization of speech processing. *Nature Reviews Neuroscience* **8**:393–402.

Hubel, D.H., and Wiesel, T.N. (1962). Receptive fields, binocular interaction, and functional architecture in the cat's visual cortex. *Journal of Physiology (London)* **160**:106–154.

Iyengar, B.K.S. (1981). *Light on Pranayama*. Allen and Unwin, London.

Jackendoff, R. (1983). *Semantics and Cognition*. MIT Press, Cambridge MA.

Jackendoff, R. (2002). *The Foundations of Language: Brain, Meaning, Grammar, Evolution*. Oxford University Press, Oxford UK.

Jakobson, R., and Waugh, E. (1979). *The Sound Shape of Language*. Harvester Press, Brighton Sussex.

Jespersen, O. (1963). *The Philosophy of Grammar*, 9th edition. Allen and Unwin, London.

Joshi, A. (1987). An introduction to tree-adjoining grammars. *In* Manaster-Ramer, A., editor. *Mathematics of Language*. J. Benjamins, Amsterdam, pp. 87–114.

Kandel, E.R. (1995). Brain and behaviour. *In* Kandel, E.R., Jessell, J.H., and Schwartz, Th.M., editors. *Essentials of Neural Science and Behavior*. Prentice Hall, London, pp. 6–19.

Kandel, E.R., Jessell, J.H., and Schwartz, Th.M., editors. (1995). *Essentials of Neural Science and Behavior*. Prentice Hall, London.

Kant, I. (1790). *Critique of Teleological Judgement.*

Karmiloff, K., and Karmiloff-Smith, A. (2002). *Pathways to Language*. Harvard University Press, Cambridge MA, p. 12 ff.

Klein, W., and Perdue, C. (1997). The Basic variety, or: couldn't language be much simpler? *Second Language Research* **13**:301–307.

Lakoff, G., and Johnson, M. (1999). *Philosophy in the Flesh – The Embodied Mind and its Challenge to Western Thought*. Basic Books, New York, p. 19.

Langacker, R.W. (2008). *Cognitive Grammar – A Basic Introduction*. Oxford University Press, Oxford.

Lashley, K. (1950). The search of the engram. *Society of Experimental Biology Symposium*. **4**:454–482.

Le Doux, J. (1996). *The Emotional Brain – Mysterious Underpinnings of Emotional Life*. Touchstone, New York.

Leibníz, G. (1714). *Monadology*.

Lichtheim, L. (1885). Über Aphasie. *Deutsches Archiv für klinische Medizin* **36**: 204–268.

Merleau-Ponty, M. (2002). *Phenomenology of Perception*. Routledge and Kegan Paul, London.

Miller, G.A., and Johnson-Laird, Ph.N. (1976). *Language and Perception*, 2nd edition. Harvard University Press, Cambridge MA.

Montague, R., and Schnelle, H. (1972). *Universale Grammatik*. Vieweg, Braunschweig.

Moscovitch, M., Chein, J.M., Talmi, D., and Cohn, M. (2007). Learning and memory. *In* Baars, B.J., and Gage, N.M., editors. *Cognition, Brain, and Consciousness*. Academic Press, London.

Mountcastle, V.B. (1998). *Perceptual Neuroscience – The Cerebral Cortex*. Harvard University Press, Cambridge MA.

Nauta, W.J.H., and Feirtag, M. (1986). *Fundamental Neuroanatomy*, Part II. W.H. Freeman, New York.

Penrose, R. (1989). *The Emperor's New Mind – Concerning Computers Minds and the Laws of Physics*. Oxford University Press, Oxford UK.

Petrides, M., and Pandya, D.N. (2002). Association pathways of the prefrontal cortex and functional observations. *In* Stuss, D.T., and Knight, R.T., editors. *Principles of the Frontal Lobe Function*. Oxford University Press, Oxford UK, pp. 31–50.

Poincaré, H. (1908). *Science et Methode*. Flammarion, Paris.

Polya, G. (1965). *Mathematical Discovery*. John Wiley and Sons, New York.

Popper, K.R., and Eccles, J.C. (1977). *The Self and its Brain*. Springer, Berlin.

Posner, M.E., and Raichle, M.E. (1997). *Images of Mind*. Scientific American Library, New York, p. 101.

Powley, T.L. (1999). Central control of autonomous functions – The organization of the autonomic nervous system. *In* Zigmond, M.J., Bloom, F.E., Landis, S.C., Roberts, J.L., and Squire L.R., editors. *Fundamental Neuroscience*. Academic Press, San Diego, CA, pp. 1027–1048.

Premack, D., and Woodruff, G. (1978). Does the chimpanzee have a theory of mind? *Behavioural and Brain Science* **1**:515–526.

Pulvermüller, F. (2002). *The Neuroscience of Language*. Cambridge University Press, Cambridge UK.

Rafal, R. (2002). Cortical control of visuo-motor reflexes. *In* Stuss, D.T., and Knight, R.T., editors. *Principles of Frontal Lobe Function*. Oxford University Press, New York, pp. 149–158.

Rescher, N. (1991). G.W. *Leibniz's Monadology – An Edition for Students*. Routledge, London.

Rizzolatti, G., Luppino, G., and Mattelli, M. (1995). Grasping movements: visuomotor transformations. *In* Arbib, M.A., editor. *The Handbook of Brain Theory and Neural Networks*. MIT Press, Cambridge MA.

Rizzolatti, G., Fogassi, L., and Gallese, V. (2000). Cortical mechanisms subserving object grasping and action recognition: a new view on the cortical motor functions. *In* Gazzaniga, M.S., editor. *The New Cognitive Neuroscience*. MIT Press, Cambridge MA.

Rizzolatti, G., and Craighero, L. (2004). The mirror neuron system. *Annual Review of Neuroscience* **27**:169–192.

Sanes, T.H, Reh, T.A., and Harris, W.A. (2006). *Development of the Nervous System*. 2nd edition. Elsevier Academic Press, Boston.

Schnelle, H. (1978). Poetische Sprache und poetischer Zustand bei Paul Valéry. *In* Schmidt-Radefeldt, J., editor. Paul Valéry. Wissenschaftliche Buchgesellschaft, Darmstadt, pp. 247–267.

Schnelle, H. (1980a). Wittgenstein on time and tense and the linguistic turn. *In Proceedings of the IVth International Wittgenstein Symposium*. Hölder – Pichler – Tempski, Wien, pp. 525–531.

Schnelle, H. (1980b). Pretense. *In* Rohrer, Chr., editor. *Time, Tense, and Quantifiers*. Niemeyer, Tübingen.

Schnelle, H. (1980c). Introductory remarks on theoretical neurolinguistics. *Language Research (Seoul)* **16**:225–236.

Schnelle, H. (1981a). Phenomenological analysis of language and its application to time and tense. *In* Parret, H., Sbisà, M., and Verscheuren, J., editors. *Possibilities and Limitations of Pragmatics*. J. Benjamins, Amsterdam, pp. 631–655.

Schnelle, H. (1981b). Introspection and the description of language use. *In* Coulmas, F., editor. *Festschrift for Native Speaker*. Mouton, The Hague, pp. 105–126.

Schnelle, H. (1981c). Semantics and pragmatics in psycho-physiological context. *In* Klein, W., and Levelt, W., editors. *Crossing the Boundaries of Linguistics*. Reidel, Dordrecht.

Schnelle, H. (1981d). Introduction and Editing; Sprache und Gehirn – *In Honor of Roman Jakobson*. Frankfurt/M: Suhrkamp.

Schnelle, H. (1981e). Elements of theoretical net-linguistics. *Theoretical Linguistics 8*:294–329.

Schnelle, H. (1983). Some preliminary remarks on net-linguistic semantics. *In* Rickheit, G., and Bock, M., editors. *Psycholinguistic Studies in Language Processing*. de Gruyter, Berlin, pp. 82–98.

Schnelle, H. (1988). Turing naturalised: Von Neumann's unfinished project. *In* Herken, R., editor. *The Universal Turing Machine – A Half-Century Survey*. Oxford University Press, Oxford UK, pp. 539–557.

Schnelle, H. (1991). *Die Natur der Sprache*. de Gruyter, Berlin.

Schnelle, H. (2004). A note on enjoying strawberries with cream, making mistakes, and other idiotic features. *In Alan Turing: Life and Legacy of a Great Thinker*. Springer, Berlin, pp. 353–58.

Schnelle-Schneyder, M. (1990). *Photographie und Wahrnehmung*. Jonas Verlag, Marburg.

Schnelle-Schneyder, M. (2003). *Sehen und Photographieren. Springer*, Berlin.

Schwartz, J.H. (1995). *In* Kandel, E.R., Jessel, J.H., and Schwartz, Th.M., editors. *Essentials of Neural Science and Behavior*. Prentice Hall, London, pp. 6–19.

Searle, J.R. (1969). *Speech Acts*. Cambridge University Press, Cambridge UK.

Searle, J.R. (1983). *Intentionality – An Essay in the Philosophy of Mind*. Cambridge University Press, Cambridge UK.

Searle, J. (1992). *The Rediscovery of the Mind*, 3rd edition. MIT Press, Cambridge MA.

Searle, J. (1998). *Mind, Language and Society*. Basic Books, New York.

Sgall, P. (2006). *In* Hajicova, E., and Panevova, J., editors. *Language in its Multifarious Aspects*. The Karolinum Press, Prague.

Singer, W., and Ricard, M. (2008). *Hirnforschung und Meditation. Ein Dialog (in German)*. Suhrkamp, Frankfurt/Main.

Stern, D.N. (1990). *Diary of a Baby*. Basic Books, New York.

Stern, D.N. (2000). *The Interpersonal World of the Infant*, 2nd edition. Basic Books, New York.

Szentagothai, J., and Arbib, M.A. (1975). The module concept in cerebral cortex architecture. *Brain Research* **95**:475–496.

Tomasello, M. (2003). *Constructing a Language – A Usage Based Theory of Language Acquisition*. Harvard University Press, Cambridge MA.

Treisman, A. (1988). Features and objects: the Fourteenth Bartlett Memorial Lecture. *Quarterly Journal of Experimental Psychology* **40A**:201–237.

Wachsmuth, I. (2008). 'I, Max' – Communicating with an artificial agent. *In* Wachsmuth, I., and Knoblich, G., editors. *Modelling Communication with Robots and Virtual Humans*. Springer, Berlin.

Wierzbicka, A. (1999). *Emotions Across Languages and Cultures*. Cambridge University Press, Cambridge UK.

Wegener, P. (1885). *Grundfragen des Sprachenlebens*. Niemeyer, Halle. (In German).

Author index

Subject index

Printed in the United States
by Baker & Taylor Publisher Services